SAIL AND RIG

THE TUNING GUIDE

MAGNE KLANN AND ØYVIND BORDAL

SAIL AND RIG
THE TUNING GUIDE

Blue Ocean Media

ORIGINAL TITLE
Seil og rigg
Håndbok i trimming

TEXT
Øyvind Bordal

PHOTOS, ILLUSTRATIONS
Magne Klann

TRANSLATION
Øyvind Bordal, Harald Bjerke

PROOFREADING
Justin Crowther

LAYOUT
Grafikhuset ApS, Inger Chamilla Schäffer

COVER PHOTOS
Magne Klann

EXTERNAL PHOTOGRAPHERS
Redningsselskapet
Daniel Forster, Volvo Ocean Race
Tobias Stoerkle, blende64.com/Bavaria

PRINT
Specialtrykkeriet Viborg
1. edition 2016

ISBN 978-82-998611-7-5

THANKS TO...

Axel Nissen-Lie
Bengt Lindholm
Bjørn Olav Tveit
Christina Wallin
Fredrik Lööf
Henrik Søderlund
Henrik Torgersen
Jørn Erik Ruud
Kjell Inge Heiberg
Magnus Mørk
Malin Bergman
Morten Ullmann
Moss Seilforening
Per Arne Skjeggestad
Per Christian Bordal
Soon Seilforening
Stein Thorstensen
Team Jake
Team Magic
Thomas Nilsson
Øystein Fuglesang

INTRODUCTION

This book has one overall ambition: To communicate an understanding of the dynamics of rig and sail that makes sailing even more attractive.

Attractive in this context means several things. There are a number of reasons why a well-trimmed sailing boat is a better place to be. The boat will heel less, balance better and sail faster – all things that contribute to the pleasure of sailing.

You will arrive at your destination sooner, increase your range and get better results when racing. Being on board will be more comfortable for the whole crew and nurture the "feeling"- this undefinable thing that is a big part of the attraction for us sailors, depending very much on the rig and sails being in harmony with the elements.

A well-trimmed boat is also safer. The risk of losing control is smaller and so is the risk of damage and technical problems. If the crew treats the rig and sails appropriately, the boat will also be more viable economically. The rig and sails are the most expensive components of the boat. All parts of the rigging will last longer and maintenance costs will be reduced when the boat is trimmed and sailed well.

This book contains lots of information in the form of text, photo and illustrations. Some of it will be mostly of interest to sailors with limited experience, but much of the material should appeal to sailors with prior solid knowledge and experience. The book treats the different aspects of trim in depth, so even sailors at a higher level can develop their competence and become even better sailors.

In other words: There is something here to be found for a very broad group of sailors. Like any other publication from Blue Ocean Media, a guiding principle has been to make even complex matters as simple as we possibly can.

CONTENTS

CHAPTER 01   7

UNDERSTANDING THE RIG

Parts of the rig 8
The capacity af a rig 11
Masthead and fractional rig. 14
Shrouds and stays 17

CHAPTER 02   23

MATERIALS IN THE RIG

Aluminium. wood and carbon 23
Materials in standing rigging 28
Materials in running rigging 30

CHAPTER 03   33

RIGGING AND BASIC TRIM

Using the mast crane 34
Rig trim – a workflow that works 36
How to measure rig tension 42
Consequences of changing rig trim 45
Going up the mast 46
How to check the rig 48

CHAPTER 04   51

HOW DO THE SAILS WORK?

Why does the boat move forwards? 52
How does lift and drag work on different points of sail? 56
Apparent wind 58

CHAPTER 05   61

THE SAILS: MATERIALS AND CONSTRUCTION

Parts of the sail 62

Materials in sails 64
How are sail profiles made? 68
Triangular or square mainsail? 70
Spinnaker and gennaker profile 71

CHAPTER 06   73
SAIL HANDLING:

Hoist, take down, furling and reefing 73
Mainsail 76
Headsail 78
Furling systems 80
Furling main 83
Furler or snuffer? 84
Reefing 87
Reefing the headsail 88
Reefing the main 90
Reefing the mainsail downwind 94
Sail maintenance and repair 96

CHAPTER 07   99
SAIL TRIM

The basics 99
What is sail trim? 101
5 key elements 102
Angle of attack 104
Heel angle 106
Twist 108
Vortex 110
Balance 112
Fullness 114

CHAPTER 08   119
PRACTICAL SAIL TRIM

Using the trim lines 121
Telltales 122
The backstay 124
Forestay sag 126
Mast-bend 127
Running backstays 128
Headsail sheet 130
Headsail halyard 132
Sheeting point 134
Barberhaul/tweaker 136
Mainsail halyard 139

Cunningham 140
Outhaul 141
Kick or boomvang 142
Traveller 144
Mainsheet 146
Mainsail trim in gusts of wind 150
Spinnaker trim features 152
Lines not used for trim 156
Spinnaker topping lift 158
Afterguy 160
Helming and sheeting 162
Trim features of the gennaker 164
Sheets, tackline, halyard and barberhaul 166

CHAPTER 09   169
TIPS AND TRICKS

Rigging and sails in a storm 170
Storm Sails 172
Storm strategy 174
Pointing ability 176
Trim guide for good pointing ability 180
Trim guide for light wind 182
Trim guide for medium wind 184
Trim guide to hard wind 186
Troubleshooting 188

INDEX   **190**

FOREWORD

FREDRIK LÖÖF

America's Cup sailor, Artemis Racing
Olympic gold 2012, Star
Olympic bronze 2008 Star
Olympic bronze 2000, Finn
Five times World Champion, Finn and Star

I have worked full time with rigging and sails since I was 18 years old. This book explains in depth how to deal with rigging and sails in the best possible way. Actually, the fundamentals for elite sailors and cruisers are not as different as one might think, although elite sailors will squeeze that last little bit out of the boat, while cruisers primarily will want a comfortable voyage from A to B. With better knowledge, regatta sailors can achieve better results and cruisers can cruise more comfortably.

As a regatta sailor, I work with small margins. I've always been obsessed with developing my masts and sails, finding the perfect balance in the boat and simply sailing as fast as possible. The most important thing for me, when achieving the right balance in the boat, is working with the major factors. A lot of sailors spend a lot of time testing things like a small difference in spreader angle, or maybe a little less rig tension. But first and foremost you have to work with the fundamentals and here I am referring primarily to the mainsheet and jibsheets. It may sound simple, but even at the highest level you must constantly focus on having them under control. Only then can you move on to evaluate the finer details.

Both elite sailors and cruisers should zoom out and gain an overall assessment in order to get the correct basic setting for the boat. Everyone should start by asking themselves: What kind of conditions do we have today? What should I focus on right now?

I also believe that it is important to make it as easy as possible for yourself. Stressful situations arise at sea sooner or later, whether sail racing or cruising. If you have a plan and some thoughts on how the sails should be trimmed, everything becomes much easier – not to mention a lot safer.

Another tip to a better understanding of how the boat should be trimmed is to test extreme trim settings. For example, try releasing the jibsheet a few centimeters and note how it affects helm pressure. Then you can sheet in a little too much and experience a more direct feel to the helm. If you experiment and get a better understanding and more knowledge, the results will follow.

One more thing: When the boat has poor speed or is uncomfortable to sail, most sailors tend to change a lot of things simultaneously. I think it is better to change one thing at a time. If you do many changes at once, you won't know why it got better or worse.

And finally: Positive communication and an understanding among the crew for how the boat is sailed, is very important. On board everything, from a doublehanded boat to a 70-foot Maxi, constant dialogue between crew members is essential to be able to sail the boat faster. Greater involvement, where the whole crew participates with ideas about how to sail the boat, gives confidence. The security increases and the overall experience just gets better.

Well, these are some of my thoughts on this book – it is a really interesting and rewarding read and will definitely help you to experience more magnificent hours, days and years at sea.

Fair winds
Fredrik Lööf

01

UNDERSTANDING THE RIG

Parts of the rig 8
The capacity af a rig 11
Masthead and fractional rig .14
Shrouds and stays 17

PARTS OF THE RIG

There are various rig configurations. Here are the most common, along with names of the different parts of the rigging.

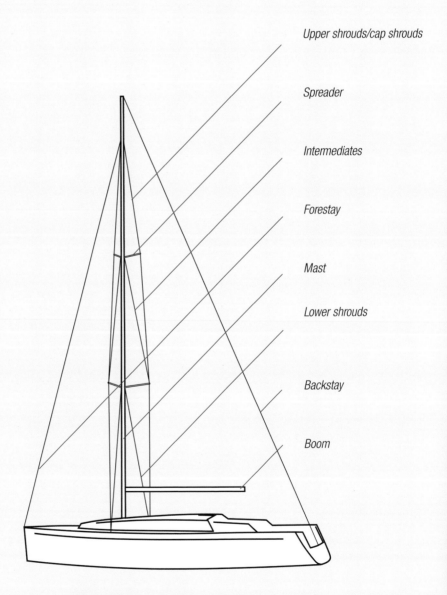

Upper shrouds/cap shrouds

Spreader

Intermediates

Forestay

Mast

Lower shrouds

Backstay

Boom

Fractional rig with swept double spreaders

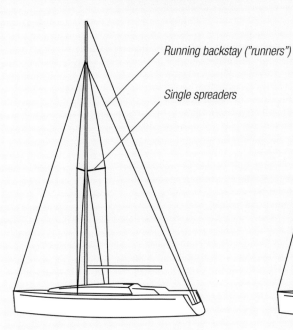

Running backstay ("runners")

Single spreaders

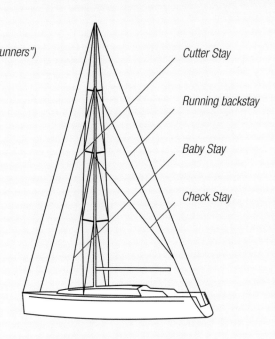

Cutter Stay

Running backstay

Baby Stay

Check Stay

Fractional rig with single swept spreaders and running backstays

Masthead rig with multiple in-line spreaders (here three sets).

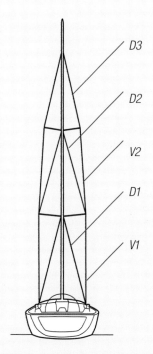

D3

D2

V2

D1

V1

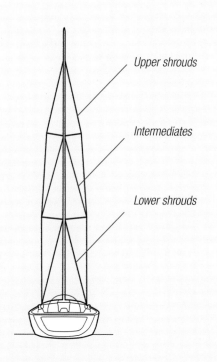

Upper shrouds

Intermediates

Lower shrouds

Riggers and professional sailors usually refer to transverse rigging using these abbreviations: V (verticals) and D (diagonals).

Here are the terms mostly used in broader circles.

THE CAPACITY OF A RIG

A rig is basically a structure that can support the sails. Closer inspection reveals that it is so much more.

At first glance a rig is a fairly simple device: A tube, usually made of aluminium, supported by steel wire usually in conjunction with transverse spreaders. Inside the tube there are lines, which can be used to hoist and lower the sails. A boom on a swivel fitting holds the rearmost corner of the mainsail out from the mast. The headsail is attached on the front wire.

CHAIN OF COMPONENTS

That's it. Actually, it's only a question of a pole that can keep the sails up.

But like everything else man made, the rig had a developmental history, with constant improvements and new ideas. A modern rig has slowly evolved into a high-tech chain of components that work together with fairly advanced dynamics, which makes it efficient, but also vulnerable. A chain is no stronger than its weakest link and here there are many of those!

MORE GAINS

Fortunately there is no need to familiarise yourself with all aspects of the construction of a rig, but there are big advantages associated with a basic understanding of the forces at work in the rig and the function of the different components. You can be better equipped to trim the rigging and sails, prevent damage and ensure a longer lifespan of what may be a substantial investment.

TENTS AND GUY ROPES

Anyone who has been on a camping trip, knows that it is best to put a rope at some distance away from the tent pole. If the peg is pushed into the soil right next to the pole, it will not support particularly well. A tent peg that is knocked down right next to a tent pole will also push the pole hard into the ground. The peg is easily pulled out because the cord must be very tight if it is to support properly. If you use a thin cord that stretches, the tent has poor support.

COMPRESSION AND TENSILE STRENGTH

We experience the same things on a sailboat. The mast is subjected to large compressive forces, because the support points for stays and shrouds are quite close to the mast foot. Consequently, the requirements for tensile strength in all components are very high.

Support in the longitudinal direction is best; boat length is often about 0.7 times the mast height. This ensures a pretty good angle on the forestay and backstay and other stays that the rig may be equipped with.

Beam support is something else. As regular keelboats are not particularly wide - usually somewhere between two and four meters – shrouds must be attached somewhere between one and two meters from the mast. Considering the length of the mast and the transverse forces it can be subjected to, it is amazing that it is possible to keep it upright at all!

DIMENSIONS AND SPECIFICATIONS

This is possible because the design and material selection are adapted to these high demands. It applies to all levels, from attachments at the top of the mast, terminals, toggles, turnbuckles and not least the chainplates and the way they are attached or integrated into the bulkhead or hull. Yes, even the hull of a sailboat is designed and built with the forces of the rig in mind. Most boats can deform several centimeters using rig tension – in fact to such an extent that interior doors can get stuck!

FACILITATE SAIL PROFILE

It's one thing to have the mast remain upright and under sail in rough weather and high seas. That is demanding enough, but sailors have even bigger ambitions. The rig should support the sails' optimum profile as well. Mast dynamics affect the sails and this influence can be detrimental. However, if the rig is trimmed correctly it will support the needs and we have to adapt the sailplan to changing conditions.

MANY GEARS AND SETTINGS

Sails are actually designed for the dynamics of a correctly trimmed rig. Together, rigging and sails constitute a driving force with many gears and settings. Most rigs can be set once and for all, so they fit your particular needs, your sails, your boat and your way of sailing. Some trim functions of the rig, however, can be adjusted underway, in order to adapt to the conditions. All in all, we are talking about an interaction that is quite complex, not to mention interesting!

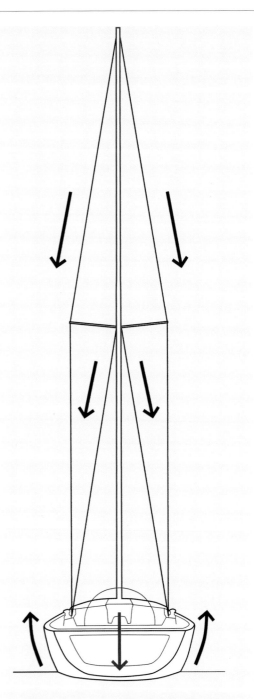

The shrouds push the mast down towards the bottom of the boat, with a great compression force. The shrouds pull in the opposite direction, pulling on the chain plates mounted in the hull. This force is also considerable. The boat has to be constructed and dimensioned to handle these conflicting loads.

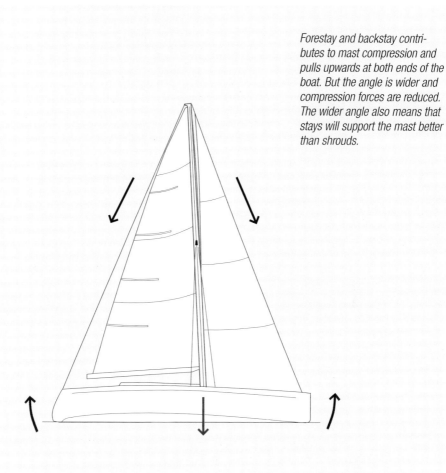

Forestay and backstay contributes to mast compression and pulls upwards at both ends of the boat. But the angle is wider and compression forces are reduced. The wider angle also means that stays will support the mast better than shrouds.

The mast will bend under compression, to some degree. The goal is to make it bend in a way that supports our needs instead of conflicting them.

Turnbuckles are used for adjusting rig tension. The deck fitting below is attached to a chain plate, attached or integrated into the hull.

WHAT EXACTLY IS THE DIFFERENCE?

MASTHEAD AND FRACTIONAL RIG

Initially it can be difficult to understand why this dividing line is of major importance, but in fact we are talking about two different types of rig, each with its own dynamics.

The difference between a masthead rig and a fractional rig is fundamentally quite simple: A masthead rig has the forestay led all the way up to the top of the mast (hence the name), while a fractional rig has the forestay fixed lower on the mast. It may be 3/4, 7/8, 9/10, or something else entirely, but a fraction of the full length the mast.

MASTHEAD RIG
In a masthead rig the backstay pulls on the opposite point as the forestay. The backstay takes hold of the forestay and tightens or slackens it. Generally, the forestay itself is not adjusted, it is tightened by the backstay pulling the masthead backwards. Mast curvature is affected to a certain extent, since downward mast-compression is increased. If one wants to bend the mast effectively in a masthead rig, however, it is necessary to have double lowers or a baby stay, which can pull the mid-section of the mast forward. A check stay running backward can control this process and stabilise the rig's mid-section. Most spreaders on masthead rigs are perpendicular to the mast (in-line). The consequence is that the backstay has to take care of the job of tensioning the forestay. The shrouds do not pull the mast back, as is the case on a rig with aft swept spreaders.

A masthead rig has an inherently larger headsail than a fractional rig. The mainsail is relatively small due to the boat being designed with a certain total sail area, partly based on stability. This area is divided between the mainsail and headsails. If one sail area increases, the other will decrease.

The masthead rig is stable and relatively simple, but usually has fewer trim possibilities than a fractional rig. Traditionally, it has been preferred by cruisers.

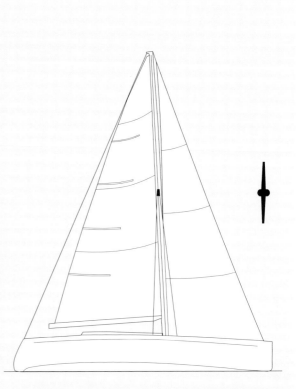

Masthead rig. The forestay mounted at the masthead. In-line spreaders.

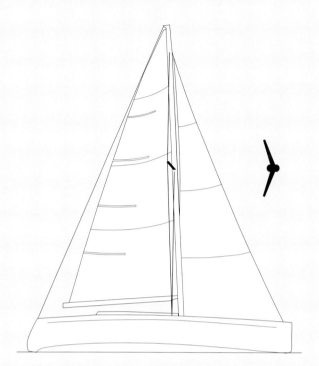

Fractional rig. The forestay mounted below the masthead.
Swept spreaders.

FRACTIONAL RIG

In a fractional rig the backstay pulls on the mast from a point that is higher than the forestay. The forestay upper attachment point pulls the mast forwards and when the backstay is tightened, you can imagine that the mast curves around this point and the compression in the mast tube will also increase. Overall, this means that the curvature propagates down the mast. The further down the mast the forestay is attached, the more effective this process becomes.

A lot of newer fractional rigs are designed with the upper section of the mast gradually becoming thinner (tapered masts). This makes it easier to bend the mast backwards, saves some weight and improves stability.

In a fractional rig you can't control forestay tension quite so easily using the backstay. Previously, running backstays were frequently used, but they are rather impractical, as they need to be set and released every time you tack or jibe. Therefore, aft swept spreaders are becoming more common. Here the shrouds pull the mast back and ensure a certain amount of pressure on the forestay, regardless of backstay tension.

A fractional rig has a smaller headsail and a relatively larger mainsail. This rig type can be trimmed more efficiently while underway and is therefore traditionally preferred by racers. Over time this picture is changing. 9/10 rigs, with aft-swept spreaders and the shrouds often attached on the outside of the hull, have become more common. This makes it possible to operate with a smaller headsail – a narrow, barely overlapping jib which can be sheeted in front of the shrouds. As the top of the mast is narrower than the rest of the mast tube, the mast curvature can easily be regulated with the aft stay. It creates an effective tool to control the forces of the rig. At the same time, the rig type is easy to handle for cruisers, being without running backstays or large cumbersome headsails. This type of rig combines the advantages of the fractional rig and masthead rig in a practical and functional way.

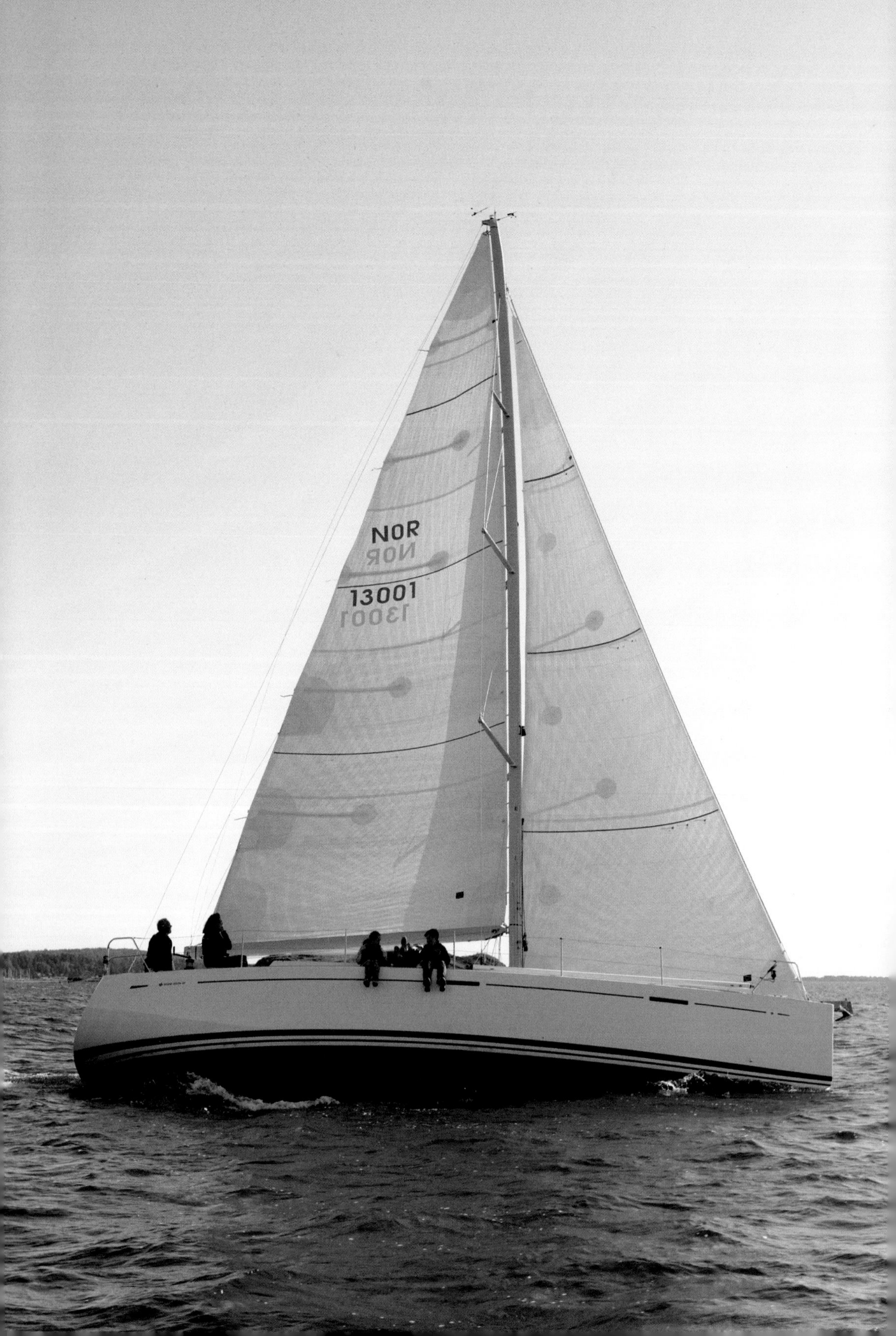
NOR
13001

SHROUDS AND STAYS

Let's examine the elements of standing rigging. We will also take the opportunity to look at what tasks the various shrouds and stays really have – and how they interact.

STANDING AND RUNNING RIGGING

"Standing rigging" is the term for all components of the rig involved in keeping the mast upright, ie shrouds, stays and there associated parts such as spreaders, terminals, turnbuckles etc.

"Running rigging" refers to everything that can run through blocks and be tightened, loosened and used to adjust sails; i.e. halyards, sheets and trim lines.

SWEPT SPREADERS

Spreaders can be mounted at 90 degrees to the mast (in-line), or at an angle aft (swept spreaders).

On a rig with swept spreaders, lowers and cap-shrouds are attached to chain plates some distance behind the mast. As a consequence, both sets of shrouds will pull the mast aft. Lower shrouds will pull on the midsection, cap shrouds will pull the masthead backwards. In this way, the cap shrouds will also tighten the forestay. In other words: On a rig with swept spreaders, the cap shrouds will create a certain permanent tension in the forestay.

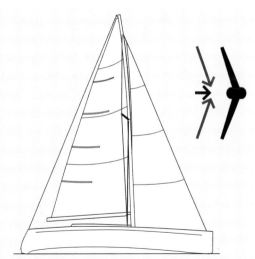

Cap shrouds push the mid section of the mast forward.

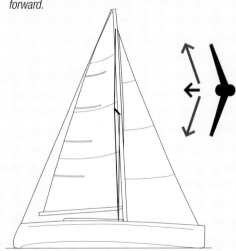

Lower shrouds pull the mid section of the mast backwards (aft).

IN-LINE SPREADERS

Spreaders perpendicular to the mast are very effective at stiffening the mast abeam, but they do not support the mast fore and aft and have no effect on the forestay. Rigs with perpendicular spreaders are very dependent on the backstay in conjunction with running backstays or check stays. Baby stays or double lowers will control the midsection of the mast.

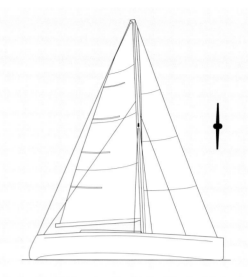

Tension in cap shrouds, lower shrouds (and intermediates, if any) will not affect longitudinal rig tension. That will only happen on rigs with swept spreaders.

CAP SHROUDS

Cap shrouds (also called uppers) runs over one or more sets of spreaders, heading toward the deck. Spreaders ensure a greater angle between the mast and shrouds and makes the whole rig far more stable sideways. It corresponds to moving tent guy lines further away from the tent pole. The result, however, is that the cap shrouds are pressing the spreaders into the mast. As there is such high tension in the shrouds, this pressure is quite substantial. When the boat is stationary, this force is countered by the opposing spreaders.

When the boat is sailing, pressure is no longer equal on both sides of the mast. Now, the tension in the shroud to windward is higher and the tension in the shroud to leeward is lower. The spreaders will push the mid-section of the mast to leeward, but the lower shrouds or intermediates will hold the mast just below the spreaders and counteract this force.

In this rather ingenious way, cap shrouds, spreaders and lowers/intermediates are stabilising both the top and middle section of the mast, keeping it straight even when the boat is heeled over from the pressure of the sails.

The cap shrouds are fixed to the mast at the same height as the forestay. They can be continuous (run unbroken all the way) or be divided into sections that run in straight

lines between the spreaders and the deck. These sections are assembled at the end of the spreaders and they perform in the exact same way as unbroken, continuous shrouds.

The tension in the cap shrouds should be very high – higher than most people imagine. It is essential to ensure that this tension is correctly set. See more on page 42.

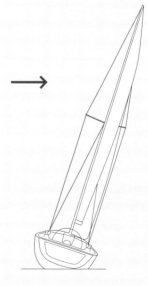

An ingenious cooperation: Cap shrouds, spreaders and lower shrouds/ intermediates are stabilising both the midsection and top of the mast. See how the forces are working when the boat is sailing.

LOWER SHROUDS AND INTERMEDIATES

The lower shrouds are mounted below the lower spreaders. They pull the midsection of the mast outward, while spreaders (using pressure from the shrouds) are pressing the midsection of the mast inward. You could say that while sailing, the windward spreader is pushing the mast to leeward, while the windward lower shrouds or intermediates are pulling the mast to windward. When there is a balance between these forces on either side of the mast, the mast becomes very stiff and strong. The same interaction manifests itself at any spreaders further up. Here, the intermediates have the same function as the lower shrouds.

When rigging a mast, this is a core factor. When all shrouds are trimmed to the required tension, the mast should be absolutely straight and centered in the boat. This should also be the case while under sail. The dynamics between cap shrouds and lowers/intermediates will ensure that the mast does not bend laterally at the spreaders.

Many boats are rigged with double lower shrouds. One is placed at an angle forwards, the other backwards. This makes it possible to use the lower shrouds to control mast curvature, while providing some support to the mast midsection longitudinally. The aft lowers on some boats are "running" i.e. they can be adjusted underway and have a function similar to a check stay.

If there is only one set of lowers, they will typically be attached to the same chainplate as the shrouds. On rigs with swept spreaders the chainplates are located aft of the mast. From this point, the lowers will also pull the midsection of the mast aft, thereby regulating the forward pressure created by the cap shrouds/spreaders on this type of rig.

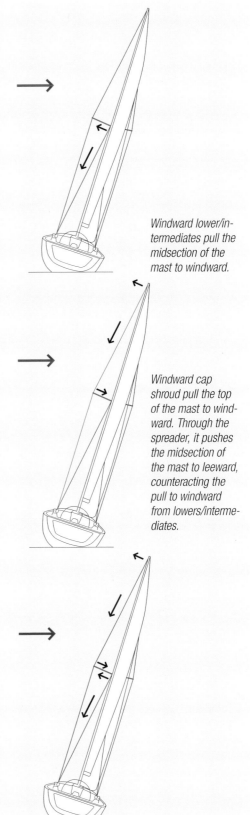

Windward lower/intermediates pull the midsection of the mast to windward.

Windward cap shroud pull the top of the mast to windward. Through the spreader, it pushes the midsection of the mast to leeward, counteracting the pull to windward from lowers/intermediates.

The interaction between uppers and lowers/intermediates ensures a mast that is effectively stabilised at each set of spreaders. At the same time, the top of the mast is held to windward. Mast compression increases with rig tension. The wind pressure in the sails will increase rig tension on the windward side and reduce it on the leeward side.

FORESTAY

Forestay tension is an essential issue – not just when it comes to finding a good, permanent setting for the rig, but also when it comes to sail trim. The luff of the headsail is attached to the forestay in the same way as the mainsail luff is attached to the mast. Under sail, when there is pressure in the sails and rigging, there is a certain "curvature" in the forestay. The midsection of the stay (and hence the sail) is stretched backwards. At the same time, the sail is pushing the forestay midsection out to leeward. This affects headsail profile. Sailmakers are of course aware of this and all headsails are designed with a certain so-called "sag". But it is important to control how much the forestay is deflected aft (and to leeward) while under sail, keeping it within the limits the sails and rigging were designed for. Sailing upwind in strong wind requires very high tension in the forestay.

Forestay tension is primarily regulated by the backstay. But it depends a bit on the rig type. If the rig has swept spreaders, shroud tension is important as well. If running backstays are available, they can also help control the tension in the forestay. Close hauled, even the main sheet pulls the masthead backwards and in some boats this may be a significant contribution to luff tension.

You don't tighten the forestay itself. The mast is pulled back in various ways and this regulates the tension of the forestay.

Forestay length is regulated only when the mast is stepped. It is adjusted until the longitudinal angle of the mast is correct, giving the boat good balance under sail. The mast should normally lean slightly aft – we will take a closer look in the section on rig tuning.

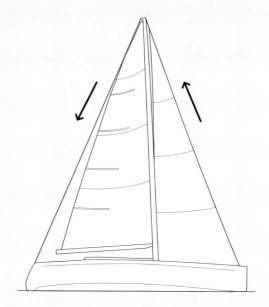

On a masthead rig (top) the backstay will tighten the forestay directly.
In a fractional rig (bottom) the backstay will also tighten the forestay, but not as efficiently. The tension will also bend the mast.

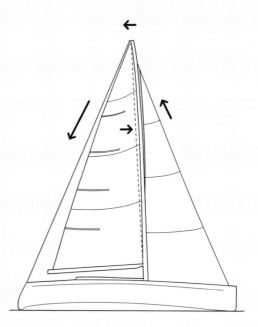

← The headsail pulls the forestay backwards and to leeward. This is called "sagg". It affects the profile of the sail.

BACKSTAY

The job of the backstay is to tighten the entire rig fore and aft. Apart from preventing the mast from falling forwards, it fulfills two other important tasks; regulating forestay tension and mast bend. Both have a great influence on the sails, we will take a closer look at this later. As we saw in the section on differences between the masthead rig and the fractional rig, the backstay has slightly different functions in these two rig types.

On a masthead rig, the backstay pulls directly on the mast opposite the forestay top attachment point, thereby regulating the forestay tension very effectively. Control of mastbend is limited, unless there are double lowers or a baby-stay that can pull the mast midsection forward.

On a fractional rig, the backstay pulls the masthead backwards, but forestay tension is not as effective as in a masthead rig. Mast-bend on the other hand can easily be regulated.

Most boats are equipped with a device that allows regulation of the backstay while sailing, although in some cruising boats, the tension of the backstay is fixed when the mast is stepped, depriving cruisers of a very effective trimming device.

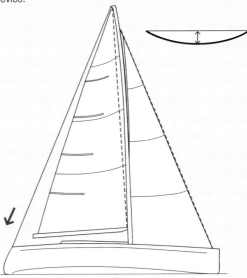

To bend the mast and straighten the forestay sufficiently you need high backstay tension. The picture shows a reasonable purchase for a 40 footer.

Tight backstay bends the mast and straightens the forestay. Both sails become flatter (above).
Loose backstay will straighten the mast and provide sag in the forestay. Both sails become deeper (below). The effect is desirable in both sails at the same time - a perfect interaction.

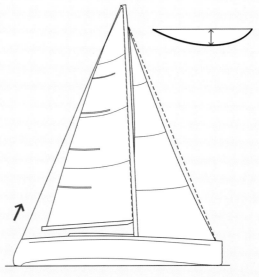

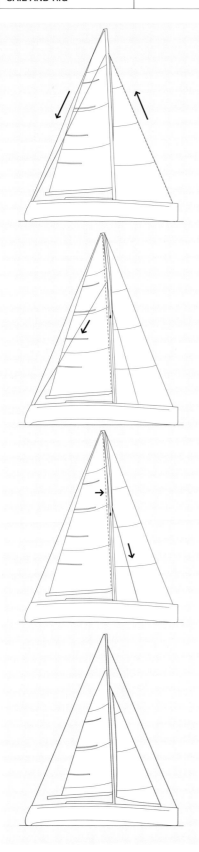

RUNNING BACKSTAYS (RUNNERS)

Running backstays are most common on fractional rigs, in order to regulate forestay tension as well as mast bend. They are very effective, as they are attached to the mast at the same level as the forestay. They also prevent the mast from "pumping" (motion fore and aft) in rough seas. They can be mounted close to the center line near the stern, where they affect the rig tension longitudinally. Mounted at the side they will to some extent affect the rig tension transversely as well. Running backstays are provided with blocks and ropes at the lower end, allowing them to be adjusted. They are only tightened on the windward side. On the leeward side they must be completely loose, in order to avoid conflict with the mainsail.

CHECK STAY

A check stay is similar to a running backstay, but does not tension the forestay. The check stay prevents the middle portion of the mast from bending too far forward. You could say that it regulates mast curvature and stabilises the mast. It also prevents the mast from "pumping" at sea. In addition, it controls the tension of an inner forestay or cutter stay, if mounted.

BABY STAY

A baby stay is used to pull the midsection of the mast forward. Combined with tension in the backstay, a masthead rig may achieve effective regulation of mast curvature. The baby stay also stabilises the rig in the longitudinal direction.

CUTTER STAY

The cutter stay is an inner forestay, where it is possible to set a staysail or cutter sail in addition to the headsail that sits on the main forestay. The cutter stay makes it possible to set two headsails, but often it is used to rig a heavy weather headsail. A cutter stay can be mounted in a way that makes it possible to rig it on and off as needed. It is usually found on bigger boats, or boats set up for blue water sailing. The cutter stay should have its own turnbuckle, to make the tension easily adjustable. Underway the tension in the cutter stay is also affected by the backstay and check stay or running back stays. The balance between cutter stay tension and forestay tension is crucial: If the cutter stay is too tight, it will be impossible to tighten up the forestay sufficiently. Attempting to do so by tightening the backstay very hard could damage the rig. If the cutter stay is too slack, it will loose effect and a possible cutter sail will not set well.

02

MATERIALS IN THE RIG

Aluminium, wood and carbon 23
Materials in standing rigging 28
Materials in running rigging 30

MATERIALS IN THE RIG

Which materials are normally used in the rig and what are the advantages and disadvantages of the different types of products available?

Shrouds and stays should not only have high tensile strength and low stretch. They must also withstand long-term stress and tolerate countless small movements and fluctuations. They also need to be adjustable and reliably secured at both ends.

In other words, we are talking about a chain of highly specialised products. On the following pages we look at the materials and types of products normally used in a modern rig.

ALUMINIUM

An overwhelming proportion of masts and spars produced in recent times are made of aluminium. The reasons are obvious: It is a light and strong material and very durable in a maritime environment. Another major advantage of aluminium is that it is very flexible -– it can withstand extensive and repetitive bending, without much deterioration. It is not a cheap material, but compared to the alternatives, it maintains an attractive price ratio. Moreover, the aluminium mast and spar industry has been around for many years, which helps keep prices under control. However, an aluminium rig is a costly thing.

Aluminium does not rust, but corrosion is something you should keep an eye on – more about that in the chapter on maintenance and repairs on page 48.

WOOD

Wooden masts and booms are primarily found on classic wooden boats. Solid wooden masts are relatively cheap to produce, but the high weight has a negative effect on the boat's stability. The forces in the rig are so great that the size of the rig must be limited, unless the boat is designed with an enormous stability and dimensioned for extreme loads (such as boats designed to the mR-rule –12-meters etc).

Today, wooden masts are usually laminated with a hollow core using epoxy. This produces a mast with far better properties, while giving the aesthetics suitable for classic sailing vessels. A properly built laminated wooden mast can withstand bending and if it is not too thick, it can be trimmed more or less like masts built of other materials.

CARBON

Carbon (also called carbon fiber) is an advantageous material in many areas on a boat, including the rig. In fact, an entire rig, including sails, standing rigging, mast and boom can be produced entirely from carbon. The reason it is so popular is pretty clear: It is a material with an extremely good correlation between weight and strength. A very light carbon structure can therefore withstand very high loads. Carbon, in common with steel and aluminium, comes in many varieties, but the strongest can be up to five times stronger than steel. It is a fibrous material, like fiberglass and molded in the same manner, usually with epoxy resins.

WEIGHT SAVING – ALSO IN THE KEEL

In a sailing boat, saving weight is a huge advantage. An improved weight to strength ratio provides better weight distribution, stability and balance in the boat. Everything can be sized down, not just because the material is stronger, but also because a light boat and rig creates less resistance and thus generates less stress and smaller loads. A sailing boat that makes use of carbon, whether in the hull or rigging (or both) will have a better speed potential. A few kilos saved in the top of the mast is of great importance to the boat's stability. Boats with carbon masts can be designed with less keel weight to the same stability, thereby reducing the boat's total weight far beyond the weight saved in the rig. In a keelboat, 10 kg weight saved in the rig normally means that about 75 kg can be removed from the keel. You can also choose to keep the keel weight and aquire greater stability. A better righting moment opens up for more sail area and more potential speed. It can, however, be worthwhile to look at the weight saving achievable by only changing the standing rigging from steel wire to carbon fiber or other fiber types. Weight saving here can in some cases be as large or larger than changing from aluminium to carbon in the mast and boom.

HIGH COST

Because carbon is a composite material, it doesn't have the well known oxidation problems that metals have, in terms of rust, corrosion, etc., although experience indicates that stainless steel can corrode when in constant contact with carbon.

Carbon can easily be manufactured to be flexible and can withstand being bent, but it is also a brittle material, with limited impact strength. Spinnaker booms hitting carbon masts in a jibe have brought down many a rig. The material's biggest drawback are the financial costs. A carbon rig is a big investment! Raw material prices are high and the production is high-tech and specialised.

FOR FAST BOATS

For boats with a higher speed potential, or boats that are primarily used for racing, a carbon rig will deliver noticeable and significant benefits. Planing boats (especially dinghies and multihulls) experience big gains. For more traditional keel boats, sailing within their theoretical hull speed and primarily used for cruising, a carbon rig will hardly deliver results that will justify the investment.

Everyone knows that carbon is strong, lightweight and expensive. But what are the practical benefits and how do you get the most out of them?

MATERIALS IN STANDING RIGGING

STEEL WIRE

When it comes to standing rigging, steel wire is the most common choice for several reasons: It is relatively inexpensive, reasonably reliable and has a good tensile strength. It is also quite easy to work with, both for riggers and boat owners. Steel wire also has the advantage that a fracture in the strands can often be spotted early. This warning of upcoming problems makes it possible to change the standing rigging before an actual failure occurs.

Most rigs are set up with 1x19 wire. One strand forms the center, six strands are twisted around this core and then the outside layers consisting of 12 further strands are twisted in the opposite direction, hence the name 1x19.

There are also other types of wire in use. Dyform is one particular variation, where triangular strands are packed into a more compact wire. This type stretches a tiny bit less then normal wire.

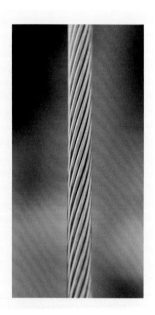

ROD

Rod is solid stainless steel, shaped into bars. It provides the maximum amount of metal for any given diameter. Rod stretches less and has greater strength than wire of the same size, which is why it is often used on racing yachts with high tension rigs. Specifications are very good, but the cost is higher than wire and riggers calculate a slightly shorter lifespan. Rod gives no warning of upcoming failure, unless small cracks are discovered at the terminals. Extra caution is needed when stepping and rigging a rod rig. Rod must not be bent too much during handling, as this may cause hidden damage and lead to subsequent breakage. That being said, rod is a proven material that has performed well for many years.

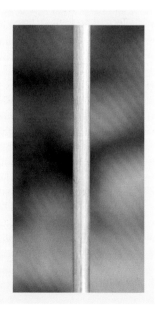

FIBER

Fiber development have resulted in many new fiber types and structures. Now it is possible to have rig solutions, where both standing and running rigging are entirely made of fiber. It can be carbon, spectra, kevlar, PBO or other types – new and enhanced products keep coming up. Common to these types of standing rigging is that they are extremely stable, extremely light and so far, extremely expensive. The dimensions can be relatively small, even for a high tension rig, so less air resistance is also an argument for choosing fiber components. Some of them are not UV resistant and must be protected from the sun. Others break easily due to impact or vibration. If you are considering using fiber it is necessary to consult professionals who can look at what specific solutions might fit your rig and your boat. In general, a fiber rig mostly makes sense on boats with high speed potential.

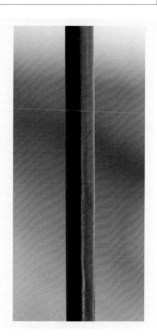

TOGGLES, RIGGING BOLTS, TURNBUCKLES AND TERMINALS

- ➕ Stainless steel quality A4 should be used wherever parts in standing rigging are connected.
- ➕ Bronze is also a good option.
- ➕ Turnbuckles or rigging screws must be lubricated and secured with a split pin.
- ➕ Bolts should fill the hole, otherwise the hole will deform and the joint will fail over time.
- ➕ Standing rigging must be subjected to linear tension, even when the mast bends under sail. This means that all types of attachments must allow the shrouds or stays to move slightly without developing kinks. Toggles are joints that facilitate this movement.

CHAINPLATES AND ATTACHMENTS

A chain is no stronger than its weakest link and standing rigging is no stronger than its attachment points. Chainplates are usually adequately dimensioned, but the chainplate attachments and the entire hull, including bulkheads, should be able to absorb the loads from the rig. It's a good idea to be familiarised with how the rig load is supported in your boat. There are many different solutions when it comes to chainplates and the way they are integrated into the boat. In some boats the rig loads are distributed to the same frame that the keel is bolted on to. In other boats shroud tension may be distributed to the main bulkhead or topsides. Rig loads are able to pull bulkheads away from the hull or break connections, especially if moisture penetration over time has weakened the structure or corroded hidden parts of chainplates or weldings. Usually there is no problem – but make sure you know how the load distribution is designed on your boat.

MATERIALS IN RUNNING RIGGING

What types of ropes are used? How do you choose the right ones?

Most tasks in a rig require ropes with as little stretch as possible. There are technically advanced ropes available, such as Spectra or Dyneema, with specifications that in certain respects are better than steel wire, able to withstand very high loads even with very small dimensions. But they are very expensive, so you have to decide whether the advantages are worth it.

IMPORTANT IN A HIGH TENSION RIG

In a carbon fiber rig with laminate sails, where everything is set up with high tension, even the slightest tendency of give in the halyard and trim lines has very noticeable negative consequences. In a rig configuration like this, it makes sense to invest in the best, most tensile ropes you can get.

GIVE ELSEWHERE?

In a wooden rig, where everything gives a little and where rig tension is low, using advanced and expensive ropes is uneconomical. The same may apply to a cruising boat with aluminium mast and older dacron sails. You do not get value for money, if the result of using a particular tensile cord results in stretch elsewhere in the rig. In many cases, a standard, pre-stretched polyester core and mantle rope will do the job just fine. Three stranded ropes or other mooring ropes however, are not for use in the rig.

WORKLOAD

When selecting rope for rig purposes, remember that breaking strength is not the same as workload. The workload, even in harsh conditions, should be far below breaking strength – perhaps as low as 30-40%. There should be a large safety margin for shock or peak loads. Breaking strength will also diminish over time.

Consider also that rope should be able to sit in a cam-cleat without slipping. Modern fiber rope is often slippery (low friction), especially when there is no mantle. If the rope is to be worked by hand, it should facilitate a good grip as well.

TYPES OF ROPES

NYLON
Nylon rope is strong, robust, soft and comfortable to handle and comes with a nice price tag as well. However it stretches far more than any other options which makes it unsuitable for use in a rig. Nylon rope should only be used for mooring and anchoring, where it works very well.

POLYESTER
Polyester rope with core and mantle has a high breaking strength and reasonably good tensile strength. This kind of rope comes in many qualities and most of them are relatively cheap. Polyester stands up well to abrasion, UV light and salt. Pre-stretched versions can be used for most tasks in the rig, but are not as tensile as the more expensive options, Dyneema, Spectra or Aramid. Polyester is often used as mantle material for more advanced types of ropes, in combination with other materials such as Cordura or Dyneema.

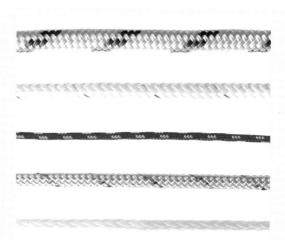

ARAMID (KEVLAR, TECHNORA, VECTRAN)
Aramid is a collective term for fibers of an older generation – especially Kevlar was popular back when there were not so many options, if you wanted rope with optimised tensile properties and high strength. These fibers generally wear down easier than Dyneema or Spectra and cannot withstand shock loads particularly well. UV tolerance is also normally not so good, so these ropes should be provided with a mantle. Breaking strength is very high, tensile strength is top class and it is also very light rope, which makes it well suited for halyards or running backstays.

Price levels and features vary from product to product, but common to them all is that they are significantly more expensive than polyester.

DYNEEMA / SPECTRA

Dyneema and Spectra are trademarks for essentially the same product: Rope in HMPE, ("High Modulus Polyethylene" Polyethylene) is chemically processed down to the molecular level and stretched in a specific direction. The result is a rope with extremely low stretch and extremely high breaking strength. It is also very robust, both in terms of UV light and abrasion. It is available with a mantle in polyester or combination of materials, but often used without. It gets a bit slippery and requires special attention when making knots or lashings and cam-cleats may have trouble holding the rope. However, durability and UV protection are top class, even without mantle. Racing boats will focus on weight and air resistance and will often choose to do so without the mantle where it is not required.

Spectra is a little lighter than Dyneema and floats, but both are subjected to "creep" - a limited, slow and permanent extension under sustained pressure. Different qualities are to varying degrees subjected to "creep". Either way you will reduce the effect by selecting rope with good margin of tensile strength relative to the load. Dyneema is a little more robust and has a slightly higher breaking strength.

Both Dyneema and Spectra can be used anywhere in the rig and being stronger than steel wire, they are particularly well suited as halyards, where the load is high and any stretch will affect sail trim. Lifelines and certain parts of standing rigging are other possible applications. Dyneema and Spectra ropes are very expensive.

03

RIGGING AND BASIC TRIM

Using the mast crane 34
Rig trim – a workflow that works 36
How to measure rig tension 42
Consequences of changing rig trim 45
Going up the mast 46
How to check the rig 48

USING THE MAST CRANE

A lot of damage can be caused by the mast crane, when the rig is stepped or unstepped. It is a job that requires forethought and planning! Here is a convenient and safe procedure.

The mast crane is potentially dangerous, not only for the mast but also for the people involved. Riggers will tell you that a lot of the damages they repair are in one way or another connected with accidents at the mast crane. Stepping or unstepping a mast can quickly go wrong! If you do not have experience with the process, it is best to get help from someone who has. A minimum of two people are required for the job and it should not be attempted in strong winds.

TAKE CARE WHEN TRANSPORTING

Transporting or moving a horizontal rig also requires caution, especially when it comes to headsail foil profile, furler system, terminals and connections. Be especially careful with rod rigging. Rod should be treated with special caution to avoid bending that may lead to subsequent rigging failure. Another area you need to keep your eye on, is the top of the mast. Most people have a Windex along with antennas, anemometer, lights and other delicate instruments that can easily be damaged – particularly by the lifting strops.

WINTER LAYUP?

If the rig has not been taken down during the winter season, you should as a minimum, hoist a person up the mast before the new season commences. It needs to be a person qualified to conduct a thorough inspection. Check and make sure that there is nothing that needs replacement or repair; you will find a checklist on page 48 of what primarily you need to look for.

For a proper inspection you need to unstep the mast and this should be done every few years. Once the rig is down you can check all terminals, connections and electrical installations.

In case of damage, especially rig failure, it may be crucial for insurance purposes that you can provide document control of the rig every few years. A professional review conducted by a professional rigger could turn out to be a very good investment.

STEPPING THE MAST

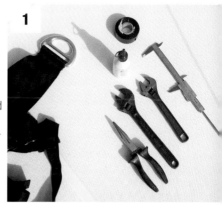

1. Make sure the necessary tools and parts to be used are close to hand and place the mast on trestles by the crane. The mast track should face down.

2. Make sure that turnbuckles, bushings, blocks and sheaths etc. are cleaned and lubricated and that everything looks OK. Are wind vanes and gear for the masthead assembled and ready? See more about maintenance on page 48.

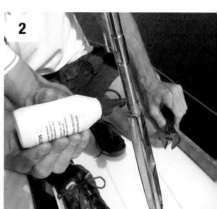

3. Find the mast pivot point. The lifting strops should sit slightly higher – about 60% of the mast length measured from the mast foot. Remember that you have be able to remove the strops again when the mast is up.

4. Prevent the lifting strops from slipping when the mast is lifted into a vertical position and make sure that the lifting hook won't damage the rig in any way. All shrouds and stays must run outside the strops. Furling systems and foils must be lashed and secured with particular care. You may want a person specifically looking after the forestay and furler while the mast is raised.

5. Moor the boat in a position where the mast crane can reach out directly above the mast step. Once your boat is positioned at the mast crane things should move pretty fast – there's likely to be a queue and it is good manners to have everything prepared in advance.

6. Lift the mast up and out and raise it to a vertical position as it comes higher. One or two crew members have to control the base of the mast and guide it onto the mast step. Keel stepped masts require particular precision, so as not to damage the boat. The mast must be lowered vertically through the deck. Ensure that you can communicate effectively with the crane operator – also from a position below deck. An "intermediary" may be necessary.

7. When the mast is positioned correctly in the mast step, fasten the forestay, top shrouds, lowers and backstay. Tighten the turnbuckles enough to ensure that the mast is stable on its own, but no more. Remove the cranes strop and hook. Do not mount the boom on the mast until you are finished with the basic trim as it will get in the way and the weight of the boom affects the mast while measuring and adjusting.

FOUR BASIC PRINCIPLES

1   Transversely, the masthead should be centered in the boat. The entire mast length should be a straight column all the way down.

2   The mast should be tilted (raked) slightly backwards – usually 1-2 degrees, but on some boats considerably more.

3   Most masts need a certain permanent longitudinal bending, so-called "pre-bend".

4   When tensioned, the rig should be subjected to a dynamic balance between the shrouds (pushing the spreaders inward) and the lowers or intermediates (pulling in the opposite direction).

RIG TRIM – A WORKFLOW THAT WORKS

There are several ways to trim a mast and some are better than others. The order of things is important to achieve a good result and to save time. Here is a practical workflow covered over 9 points.

1   GET THE MASTHEAD CENTERED

Hand-tighten the shrouds and stays. Attach the backstay, but do not tighten. The first task is to get the masthead centered abeam in the boat. Measure with a halyard (or preferably a metal measuring tape, which provides greater accuracy) to a fixed point either side of the boat. Tighten the shroud turnbuckles alternately, until you measure the exact same length with the halyard or measuring tape on both sides of the boat. To avoid tensioning the rig to any extent (this will be done later) it may be necessary to slacken one side while tightening the other. Initially, it should only be hand tight. Continue until the distance between your fixed measuring point on the railing and the masthead is exactly the same on both sides of the boat.

Exactly the same measurement on both sides.

2   STRAIGHTEN MAST IN THE TRANSVERSE DIRECTION

The next task is to straighten the mast sideways all the way down to the base. It sounds simple, but this is usually the most demanding task. The mast should stand as a straight column transversely, in its entire length. The lower shrouds are tightened first – alternately, until the mast at the bottom spreader is aligned with the masthead. Intermediates can then be adjusted accordingly – in most cases their setting can be kept from the previous rig trim. Work from the base and progress upwards, hand tightening only. Keel-stepped masts are still not wedged in the deck collar at this point.

Loosen cap shroud on starboard side.
Tighten cap shroud on port side.

Loosen cap shroud on port side.
Tighten cap shroud on starboard side.

Loosen lower shroud on port side.
Tighten lower shroud on starboard side.

Loosen lower shroud on starboard side.
Tighten lower shroud on port side.

Loosen intermediate on port side.
Tighten intermediate on starboard side.

Loosen intermediate on starboard side.
Tighten intermediate on port side.

3 FIND THE RIGHT MAST RAKE

The next job is to give the mast the correct longitudinal angle. As a rule, a mast should lean aft between 1 and 2 degrees, on some boats a fraction more. The forestay must of course be fairly tight while this is measured. The main halyard can be used for measuring – a little give in the halyard is ok in this operation. Attach a heavy weight object, just above deck level. If necessary, place a bucket of water behind the mast, submerging the weight. This dampens fluctuations and motion, making it easier to measure accurately. The distance of the halyard from the mast will indicate mast angle astern (rake).

Fractional rigs should be trimmed slightly more aft than masthead rigs. 1% inclination corresponds to 1.75 cm per meter, while 3% corresponds to 5.25 cm per meter. Multiply these figures with the mast height, in order to get usable measurements for your own mast. Measure the distance between the mast track and the front of the halyard at deck level. Remember to subtract any distance between halyard and mast, created by the sheave at the masthead.

Adjust by regulating the forestay until you are satisfied. Of course, the mast should not be bent by backstay tension during these measurements. If the forestay is not set up with a turnbuckle at the bow, the length can be adjusted by changing toggles at the lower attachment point.

4 KEEL STEPPED MAST: INSERT WEDGES IN THE DECK RING.

The next task applies only to keel stepped mast and not to deck stepped rigs. Now the mast should be centered and longitudinal mast inclination is satisfactory. Now it is time to fix the mast securely in the deck ring using rubber wedges. The mast plate and deck ring are not always placed exactly in the middle of the boat, so this should not be done before the mast is centered and trimmed both longitudinally and transversely. Otherwise you may experience an S-shape in the mast which will be impossible to correct later on. The mast should be chocked /wedged in all directions without altering the mast placement in the deck ring. If need be, the mast can be pulled on at deck level with a sheet led to a winch, in order to assist with the wedges insertion. When this completed, the waterproof collar can be mounted.

5 SET THE RIG TENSION

Now is the time to set the transverse rigging to correct tension. Use the exact same number of turns on the turnbuckles on both sides, to make sure the basic trim that has already been achieved, is not lost. First the uppers, then the lowers and then the intermediates if necessary. The uppers should normally be a bit tighter than the lowers and intermediates, especially on rigs with swept spreaders. With in-line spreaders, lowers should be tighter. This will prevent a sideways mast curve, where the midsection falls out to leeward more than the top.

See page 42 for information on how you can measure the correct rig tension.

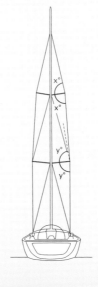

Make sure the spreaders are not pulled down during rig tensioning. Spreaders should actually point upwards a few degrees, to divide the angle between the upper and lower part of the cap shroud exactly in two.

6   CHECK MAST PRE-BEND

The next job is to control the permanent curvature of the mast – the so-called "prebend". Most masts are set with a certain longitudinal bend. The mid section of the mast (at the spreaders) should in other words curve slightly forward. This is mostly for sail efficiency, but this also affects rig safety.

On rigs with swept spreaders, prebend is determined by the relationship between the uppers and lowers. High tension in the uppers and low tension in the lowers will bend the mast. Tensioned lowers will give a straighter mast.

On masthead rigs, prebend is determined by the relationship between the aft and forward lowers and the backstay – or between the baby stay and backstay, if so equipped. Both of these scenarios require a fairly high minimum tension in the backstay, if the mast is to bend (and the forestay kept taut). A masthead rig is usually set with less prebend than a fractional rig.

Prebend prevents the rig from inverting. If the masthead is pressed forward during downwind sailing, there may be a risk that the midsection of the mast suddenly bends backwards. This can lead to rig failure. A small positive mast curvature, where the midsection is pulled forward and the masthead is held back, prevents this from happening. The risk of inverting the mast is higher on rigs with inline spreaders. Baby stays and forward lowers are consequently important tools.

Prebend may be between a quarter and a half of the width of the mast profile. Fractional rigs are often given a little more.

Now is the time to tighten the backstay, but only to minimum rig tension (see page 42). Take the opportunity to check that the mast maintains proper alignment and trim, now that rig tension is set in all directions.

Stretch the mainsail halyard down the rear of the mast and sight up along it, to see how much the mast bends.

Under special circumstances some boats may experience an advantage, if they trim the mast without any prebend, but that is under special circumstances.

7 CHECK TRANSVERSE RIG TENSION UNDER SAIL

The final parts of the process take place out on the water. Head out and sail close hauled. The boat should preferably heel 20 degrees, loading the rig to the upper end of the spectrum. Check the shrouds on the leeward side. They should feel looser – more precisely, they should begin to loose tension and be able to be moved by hand, but without hanging loose. Adjust to the required tension. This is an important process: Too little transverse rig tension increases the risk of rig failure for several reasons. As the top of the mast falls out to leeward, it creates a sharper angle between the uppers and the mast itself. This has two consequences: The uppers could become overloaded and compression forces in the mast increase radically. Too little rig tension will also allow the mast to pump while sailing in waves and this will provide countless small jerks and peak loads in all parts of the rigging.

Besides increasing the risk of rig failure, sailing performance is also significantly reduced.

When fine-tuning rig tension under sail, it is important to count the turns on the turnbuckles and do exactly the same on both sides. This way, you preserve the base trim and mast alignment. Sail both tacks and adjust the tension in the shrouds until it is satisfactory.

8 CHECK LONGITUDINAL RIG TENSION UNDER SAIL

Next step is forestay tension – in other words, longitudinal rig tension. While still sailing upwind under full pressure, sight up along the forestay. Does it fall out (sag) backwards/to leeward more than desired? Sag should fit with reasonable sail trim; in other words fit the luff curve of the headsail (more on this later). Upwind in a breeze, sag should normally be reduced to a minimum. This requires fairly high tension in the backstay. Most boats are equipped with a hauling device that enables the backstay to be adjusted underway, but some cruising boats are sailing with a fixed backstay tension. These boats should of course be extra careful to ensure that they are set up correctly during the initial rig trim.

If you have tensioned the backstay very hard to optimise upwind performance, it is important to release it afterwards. It is not healthy for a rig to have the backstay in high tension permanently.

9 CHECK ALL PARAMETERS AND SECURE THE RIG

Now, basic rig trim is complete and the rig has been test-sailed and fine-tuned fully loaded, check one last time that the mast is still aligned. Any misalignments can normally be adjusted with the lowers/intermediates. Sight up along the mast and use common sense when making adjustments. If the mast has a slight S-shape in the upper part under sail, it is usually because the intermediates are not totally symmetrical. If the masthead falls out to leeward and shrouds are tensioned, it may be necessary to loosen the top set of intermediates. Check also to make sure that longitudinal mast trim still looks right under sail.

If everything looks sensible now, rig trim is complete. Remember to lock all rigging screws with split pins and tape them to avoid damage to the sails, ropes (or crew). Use new split pins if possible and avoid G-rings – they can easily be pulled out by sheets or other moving gear. After a few days of sailing in a breeze, it is a good idea to check if there is need for re-tensioning. Often the boat gives slightly as the hull is rarely as rigid and tensile as the standing rigging.

Keep an extra eye on the mast when the mainsail is reefed. A reefed mainsail changes the dynamics of the mast and could affect its trim, both transversely and longitudinally. The midsection of the mast should not fall away to leeward. If that happens, the lowers and/or intermediates need further adjustment. Be sure to keep the backstay tight enough to prevent the mast from inverting (bending backwards at midsection), when the mainsail is reefed.

HOW TO MEASURE RIG TENSION

The "folding rule method" makes it possible to measure rig tension without expensive instruments or professional help. You need to know a few numerical values – and have a folding rule and some tape.

There are various instruments developed for measuring rig tension. It is also possible to ask a professional rigger about the ideal rig tension for your particular boat. The rigger should explain how to arrive at the correct tension values and give you tips on how to trim and take care of your rig in general.

Alternatively. here is a "do-it-yourself" method which can provide a fairly accurate measurement using tools you probably already have.

First we need a few numerical values. Shrouds and stays should generally not be pre-tensioned more than 15-20% of their breaking load – maximum 25%. The same applies for backstays, although these do tolerate 30% max load when required. Note that in a masthead rig this will result in an even higher load on the forestay (around 40%, due to the differences in the angles on the mast), so be careful. In a fractional rig, the backstay tension will not transfer directly to the forestay to the same extent. Here, you will experience more mast bend instead. In a fractional rig, you can use the backstay more aggressively (and with good results!).

Fortunately, it is possible to control how close you are to the breaking load and measure rig tension at the same time. This method is carried out by measuring how much your standing rigging stretches under load. Both wire and rod will stretch to a certain extent, as the tension approaches the breaking load of the material. This extension is predictable, constant and most important: measurable.

Here this method is described in five easy steps.

Fix the top of the folding rule to the shroud with a piece of tape.

When measuring starts, the bottom end of the folding rule must be 3-4 mm below the top of the terminal.

When tension is right, the bottom end of the folding rule is exactly level with the top of the terminal.

"THE FOLDING RULE METHOD"

1. Line up a folding rule fixed along an unloaded (turnbuckles only hand tightened) stay or shroud. Two meters, (standard length of a folding rule) is a good length to use for the measurement. Tape the folding rule to the stay/shroud by the top end only. The bottom end must extend 3-4 mm below the top of the terminal on the stay/shroud. This is now used as the measurement reference point.

_____

2. Start tightening the turnbuckle. As tension builds, the ruler begins to rise, corresponding to the wire extension over two meters.

_____

3. When you measure rigs tension, pull the ruler close to the stay/shroud. You can now read the exact distance between the top of the terminal and the end of the ruler. Measured over two meters a 1 mm extension corresponds to 5% of the wires breaking strength. This actually applies for any diameter or size of wire. Rod extends 0.7 mm per 5% breaking strength and the corresponding figure for Dyform is 0.95 mm. These figures are only valid when using a two meter folding rule.

_____

4. Keep increasing tension until you reach an extension of 1.5 mm (steel wire). Remember to count the number of turns on the turnbuckle. Then tighten the shrouds on the opposite side equally and measure again. Now the length of the extension should be doubled. When you can measure an extension of 3 mm, the shrouds or stays are pre-loaded to 15% of their breaking load (3×5 = 15).

_____

5. Normally 3-4 mm extension over two meters (15-20% of breaking load) indicates appropriate tension for cap shrouds made from steel wire. For rod the ideal extension would be barely a millimeter less. Shroud tension should be somewhat higher in fractional rigs with swept spreaders than in masthead rigs with in-line spreaders. You will only have to measure one side, as long as you take care to tighten up equally on both sides. Opposite shrouds will always have the same tension, when the boat is at rest. Remember to check that the top of the mast remains central to the boat, in the transverse direction.

_____

6. It's a wise to measure backstay tension as well. Measure to 15% minimum and 30% maximum, respectively. Mark these two points on the backstay tensioner. When trimming the rig to different conditions later on you can see what range you have to work within.

*Commonly, one design
classes have developed
very specific trim guides,
adapted to their specific
class of boat and rig.
These are usually available
online and active sailors
share them in the relevant
community.*

CONSEQUENCES OF CHANGING RIG TRIM

It is possible to optimise rig trim for extra performance in changing conditions. Sailors racing one design boats often share a special interest for this and develop very specific knowledge.

What can you really achieve if you change rig tension to adapt the sails to the actual conditions on the water? As we have seen so far in this chapter, you can achieve a solid and functional all round rig trim using the methods described. It works reasonably well in all conditions and at all wind angles. The rig will be safe, which is the most important principle number one and it provides the basis for good sail trim, which is principle number two.

DIFFERENT RIG TYPES, DIFFERENT CONSEQUENCES
Of course there are variations within this framework. You can achieve certain "special effects" by playing with tension in shrouds and stays and thereby altering their interaction, as conditions on the water change. But the dynamics of a rig under sail is relatively complex and when you change one thing, it will usually have several consequences. Furthermore, the same change does not necessarily have the same consequences in a fractional rig as in a masthead rig. This also applies to rigs with swept spreaders versus in-line spreaders, rigs with or without a baby stay, running backstays, double lowers and so on.

AN ART FORM
For ambitious racers rig trim is an art form in itself, defined by their own understanding and philosophies. Class or one design boats in particular have their own subtle trim guides with specific adjustments to the nearest millimeter for how the rig should be trimmed in different conditions and for different types of sailing. These trim guides are often available online, but top sailors in the class will also be helpful, if you are new to the community and want to join in.

Optimised rig trim is often about improving speed and pointing ability upwind in various wind conditions and (rig permitting) to adjust to the different circumstances that occur when sailing downwind.

HOW ARE SAILING CHARACTERISTICS AFFECTED?
As you can see, this is not an exact science. You can predict physical changes in the rig – but how does that actually affect sailing performance?

This is of course proves to be the most interesting part and will often turn out differently in different types of boats and is consequently outside the scope of this book.

GOING UP THE MAST

Good routines and thorough consideration will reduce the risk of accidents, when going up the mast.

Most people who are not accustomed to it will find it a little scary to climb the mast – as they should! An error or equipment failure can actually cost you your life. Focus and thorough safety procedures are essential.

DO YOUR BIT
It is a hard job to winch a person all the way to the top of the mast. Two people can do it, but three are better. An additional backup line is good practice.

To assist the person on the winch, the person going up the mast can assist by climbing – use hands and feet on the mast as well as standing or running rigging to help. Any upward effort will lighten the load on the winch. There is special climbing equipment available, but using it will require training and special knowledge.

USE COCKPIT WINCHES
The person or persons operating the winch is also at risk – that is, if a winch on the mast is used. A wrench falling from 14 meters can be deadly. It is therefore safer to use cockpit winches. From the cockpit it is also easier to see and communicate what is happening up the mast and communication will be easier.

You should of course use halyards you can trust and two are better than one. Secure the halyard directly to the harness or bosun's chair with a bowline, eliminating one source of error. A shackle or splice may fail (particularly quick release shackles). Never go up on a halyard spliced with wire. The splice usually ends up inside the mast.

BRING THE RIGHT EQUIPMENT
It is a good idea to bring a thin line for hoisting stuff up or down. An extra halyard can also be used, if available. A soft, deep bag for tools and spare parts is useful. Tie tools to your harness or the tool bag on suitable lengths of thin line to help avoid dropping them from height. Review the work to be done in advance while still on deck. This way you will know what to do, how it should be done and what you need in the way of equipment and tools.

SECURE YOURSELF WHILE WORKING
In order to minimise swinging and reduce the risk of injury, you should be secured to the mast with a separate line as soon as you reach the place you will be working. This will relieve the pressure on the halyard you are hanging on and will reduce the risk of the halyard detaching from the winch or slipping in the rope clutch down below.

DOUBLE SECURED LINES IN THE COCKPIT
The person or persons in charge of the winches carry a big responsibility. They must be alert and diligent. When the person in the mast is hoisted up, the halyards must be secured with rope clutches, cleats, or similar devices, as well as on the winch itself. Make sure to avoid riding turns on the winch.

COMMUNICATION WITH SIGNALS?
From the top of the mast it can be difficult to talk with people on deck, so it may be necessary with some simple signs. For example, you can knock on the mast with a tool; once for stop, twice for down and three knocks for up. A portable VHF or a mobile phone can also be used in tall masts or noisy conditions.

UP THE MAST UNDERWAY?
You should go to great lengths to avoid having to hoist a person up the mast while the boat is under sail. If this is absolutely necessary, be sure to reduce roll and pitch as much as possible first. Reduce sail or sail in a more favorable angle to the wind and waves. Unless it is completely flat water and light wind, it is difficult and very hard work to hold on, because the motion of the boat is amplified many times up the mast. If the boat heels, try going up along the cap shroud on the leeward side, or secure yourself with a shackle running on a tight halyard along the mast. Going up inside of the mainsail (on the windward side) is also an option. But in any case, this is a dangerous job and should only be performed in emergency situations – or by professional sailors with special experience. A helmet is always a good idea.

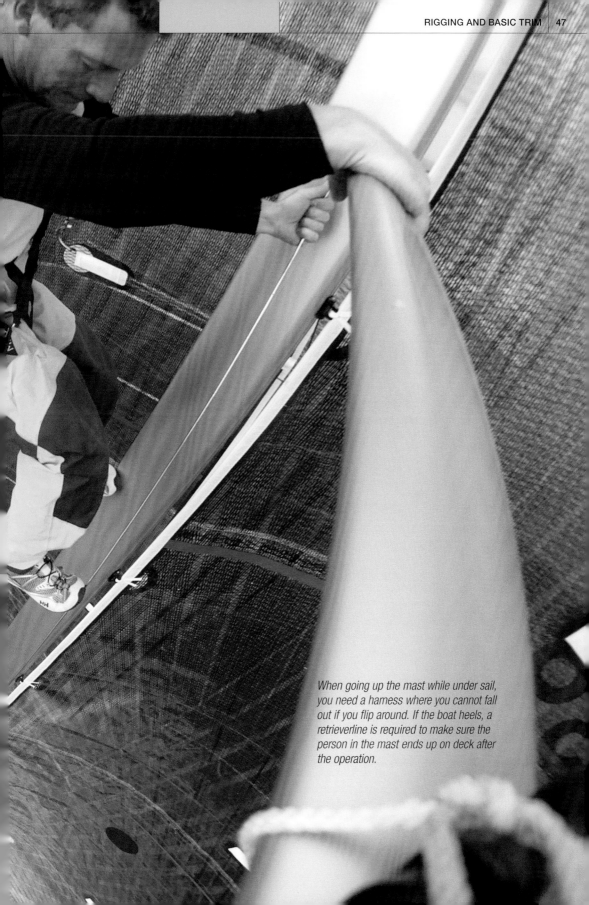

When going up the mast while under sail, you need a harness where you cannot fall out if you flip around. If the boat heels, a retrieverline is required to make sure the person in the mast ends up on deck after the operation.

HOW TO CHECK THE RIG

Thorough inspection of the rig prolongs its life and prevents expensive damage. Here you can see how it is done.

In order to carry out a proper rig check, the mast must be taken down. Once the mast is down, shrouds and stays can be dismantled. Only then is it possible to see what is hiding inside joints and connections and other places that are otherwise obscured. It is also easier to detect small cracks or corrosion in terminals and T-bar swages. You can now clean and lubricate in places more or less impossible to get to when the rig is stepped and tensioned. You have access to lines, halyards and cables running inside the mast. The top of the mast with all its small fittings for antennas and other equipment is also very difficult to access when the mast is up.

RIG INSPECTION OF STANDING MAST

There is no need to bring down the mast every winter and many people allow the mast to stand for several years. The advantage (aside from the financial savings) is that you avoid the risk of damage when the boat is rigged and derigged. But if the rig has been up since last season, it is necessary to check it over from top to bottom. There is a description of how to safely get up the mast on page 36. It is a good idea to start at the top and work your way down. Here is a brief job description:

ARE RIG BOLTS SECURED?

Make sure that all rig bolts are secured with a split pin or ring. A rig bolt has to fill the hole completely, otherwise it will deform the hole over time. A rig bolt that is undersized can also be bent easily. At worst, such a bolt can be pulled out of the hole, or crack. Do not substitute a rig bolt with a conventional threaded bolt as threads act as a file and will destroy the hole.

Take the opportunity to verify that all sheaves run freely and that the shafts and the edges of the sheaves are not damaged.

AIM ALONG SHROUDS AND STAYS

All rig terminals must be inspected for small cracks and corrosion. Aim your sight along the wires, making sure they run straight. If there is a slight angle at the ends (the attachment on the mast or at the other end, the turnbuckle) it is necessary to loosen the rig tension and correct this.

REMOVE DIRT AND MOISTURE

Check for dents or damage to the mast profile. It does not take much deformation before the mast is in danger of collapsing under the compression that occurs while under sail.

If there is a protective cover on the end of the spreaders, it should be removed in order to inspect underneath. Remove any moisture, dirt or salt. It is actually a good idea to wash the entire rig.

DO NOT STRAIGHTEN A BENT TURNBUCKLE

Boom and kick-fittings are subjected to shock loads which could damage or deform them. If that has happened, they must be replaced.

Also check the turnbuckles and lubricate their threads. Turnbuckles should be straight and the threads must be undamaged. If this is not the case the turnbuckle should be replaced. A bent turnbuckle can not be straightened – the steel becomes brittle.

WATER UNDER DECK FITTINGS?

It is also important to check chainplates and deck fittings. Water can easily penetrate under deck fittings and corrosion can develop unnoticed. Water can also penetrate along chainplates and weaken bulkheads or other attachment points for chainplates. If there is evidence of water intrusion, the fittings must be taken off, cleaned, sealed and refitted. If bolts or fittings are rusted or corroded, they must be replaced. Also, inspect welds in the deck fittings, both above and below deck.

INSPECTION OF STANDING RIGGING

A broken strand means that the wire has to be replaced. You may want to consider whether it is time to replace the entire standing rigging. As a minimum the opposing shroud should be changed at the same time. Standing rigging has a lifespan that is difficult to assess. The life expectancy is more dependent on how the rig has been treated than how old it is. Most riggers recommend replacing standing rigging after 15 to 20 years. Rod rigging should be changed more frequently. If the boat has sailed a lot, crossing oceans or blue water cruising, standing rigging should be changed with shorter intervals.

ROLLER FURLERS

Although roller furlers are reasonably robust and do not require much maintenance, they should be inspected from time to time. It is a good idea to wash and rinse the both top and bottom with fresh water and a little detergant. Some roller systems are designed for lubrication with Teflon, others are water-lubricated. It is important to know if and where the furling system on your boat should be lubricated and if so, how. Check for loose nuts and bolts and inspect the aluminium profile, especially the joints. Take the opportunity to inspect the roller line.

MOVE THE HALYARD CHAFE POINTS

Halyards are often worn on a few specific points, typically where they are locked in the rope clutch, where they run into the mast and especially where they exit the mast at the top. A good tip is to move chafe-points every year by cutting off a small piece at the end of the halyard (at the sail end). You can also turn the halyard around and make it run in the opposite direction. If you want to be extra thorough, take the halyards out of the mast when derigging at the end of the season. Leave a thin pilot line through the mast, so the halyards can easily be pulled back into place again. While the halyards are out, they can be washed and cleaned from dirt and salt. If the halyard is spliced with wire, get the splice checked while you have the chance.

CORROSION

Most often, corrosion in aluminium occurs when in direct contact with stainless steel. It also occurs where there is moisture or water (especially if salt is present) and where damaged electrical cords are touching the aluminium. If the rig is painted or varnished, corrosion can easily develop under the paint. Bubbles in the paint is a clear sign.

The top of the mast should be sealed, to prevent or at least minimise rainwater penetrating the mast tube. If necessary, fittings and bolts should be removed and resealed. Inspecting the wiring is important – insulation must not be chafed. Any electrical current in touch with aluminium will trigger corrosion.

The base of the mast can also become corroded, especially if the mast is keel stepped. The mast step area can easily collect rainwater or saltwater. Electrical wires also exit here and should be checked for damage.

Moderate surface corrosion is not dangerous. Aluminium is a material that protects itself well by oxidising. However, severe corrosion that is not detected can over time dissolve aluminium, reducing material thickness which can ultimately provoke rig failure.

A rig usually fails as a result of a broken stay, shroud or spreader and it very rarely breaks in the middle. The failure typically occurs at one of the ends, either inside the pressed terminal or at the attachment point in the mast. If allowed to accumulate in these places, dirt and salt water will create favorable conditions for corrosion. Wash, rinse , dry and inspect these core points every year! White powder is a sure sign of corrosion.

4 TIPS FOR RIG INSPECTION

1 Read the roller furling service manual and service the system according to the manufacturers instructions.

2 Check both ends of all shrouds and stays for corrosion, small cracks or faults in welds. Poor quality terminals and joints with fine cracks are some of the most common causes of rig failure.

3 Stop corrosion. Insulate stainless steel from aluminium, minimise water intrusion in the mast and make sure there is drainage at the foot of the mast. Prevent stray currents from damaged wires.

4 Use professional help if you lack experience or knowledge yourself, especially if you have an older rig with an unknown past.

04

HOW DO THE SAILS WORK?

Why does the boat move forwards? 52
How does lift and drag work on different points of sail? 56
Apparent wind 58

WHY DOES THE BOAT MOVE FORWARDS?

It is perfectly possible to sail without understanding how a sail works. But it is a lot easier to trim sails and rigging correctly, with some understanding of what actually makes the boat move. It also makes it more fun.

Can you sail faster than the wind? Yes, this Flying Phantom can move three times faster than the wind. This can be achieved because most of the wind that blows over the sails, is induced wind created by the boat itself, using speed. The more air that flows over the sails and is deflected by the sail profile, the more forward thrust is created. Foils are lifting this catamaran out of the water at a certain speed, when this occurs there is very little resistance to limit further acceleration.

Most more traditional sail boats than this, create considerably more resistance than this and haveh a limited hull speed. But even though we have to settle for more moderate speeds, we can still sail with fun and speed! In this chapter we will look at why. In other words: How does a sail actually work?

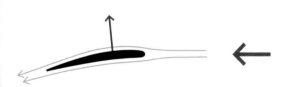

Cross section of an airplane wing illustrating lift

LIFT AND DRAG

An airplane wing pushes air down. When the air is pressed down, the wing is pushed up. In other words, the wing is lifted up by the opposite force that presses the air down. Both aircraft wings and sails force airflow to change direction and that is precisely what makes both of them work. Try putting your arm out of a car window while the car is moving and hold your hand at a slight angle upward. Your hand will be pushed pretty strongly upwards!

Actually, it is both sides of the wing profile that creates this lifting force. The underside of an airplane wing corresponds to the weather side of a sail and the upper side corresponds to the leeward side of a sail. The force has quite a logical name, used in sailing worldwide: "Lift".

There is also a lot of resistance created when a wing profile travels through an airflow. If you think of your hand sticking out of the car window again, you will quickly notice that the hand is forced backwards as well as upwards. This resistance also has a logical name: "Drag".

MAXIMISE LIFT, MINIMISE DRAG

If a sailboat is to move forward, the "lift" component must produce a pull forwards, even after the "drag" component is included. On a sailing boat, sails and rigging are designed and built to do just that. an important part of this job is done under water, in cooperation with the keel and rudder, an important part of the job is done under water. But to make it work, especially upwind, the crew has to trim and adjust the sails and thereby control the lift and drag forces that are produced. The quality of trim will decide how effectively the boat will move forward.

THE SECRET ON THE LEEWARD SIDE

There is an invisible force behind the sail that really makes the difference.

As we have seen, a sail works when the airflow is deflected. However, if the sail is to be able to pull a boat into the wind, or function with any real efficiency, this force has to be refined and fully utilised. This is where the leeward side comes into the picture.

PRESSURE DIFFERENCE

In order to create an effective force in a sail, a "laminar airflow" has to be created. This is an air flow moving over both sides of the sail, following the surface all the way from the entry to the exit points. When this happens, a difference of pressure is created between the windward and leeward side. Lower pressure on the leeward side will force the wing profile (in our case the sail) to leeward. Assuming that the sail has a certain angle to the centerline of the boat, the sail will now pull the boat forwards.

STRONG EFFECT ON THE LEEWARD SIDE

It is easy to see how airflow on the windward side is compressed and deflected by the sail. It simply blocks the wind at a certain angle, forcing it to change direction. It does this pretty well, no matter how the profile is designed. But things become a little more complicated when we look at what happens on the leeward side. This is where much of the secret is hidden. The leeward side of the sail will actually deflect the wind too, even though the airflow is not physically blocked. But this will only happen with laminar air flow and that requires a properly designed profile – and a certain angle to the wind. If operated correctly, the lift produced on the leeward side can actually be stronger than the lift produced on the windward side. So, if you want to create a truly effective lift – producing airfoil, the wind also has to stay laminated on the leeward side of the sail. Many years of research and development have made it possible to define a number of optimal profiles for exactly this purpose.

STALLING

An air flow (or water flow for that matter) tends to follow a profiled curve and will be deflected to a certain point. If you hold a spoon vertically under an open water tap, the water flow curves and follows around the back of the spoon. The scientific term for this is the "Coanda effect".

If the angle of attack becomes too large, however, or if the profile is not properly designed, the flow will lose hold and turbulence will occur, first on the leeward (low pressure) side. When that happens, the flow is no longer laminar and the lifting force is lost. The academic, international term for this is "stalling". This can also happen with airplanes and with a rather serious consequence: If the wings of an airplane stall, it will fall out of the sky.

Sail trim is all about creating a profile in the sail that can keep laminar airflow at a maximum angle against the wind and thus create a maximum deflection, without the airflow stalling.

Wind

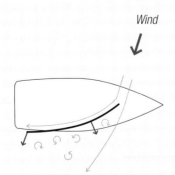

Stalling is what happens when the angle of attack becomes too big and the airflow separates on the leeward side and becomes turbulent. In a sailboat, it is not always so easy to tell when that happens.

DEEPER SAIL, MORE LIFT

Lift in a sail will increasingly develop as the sail profile becomes deeper. In other words: Deeper sail, more lift. This happens until a certain point, when the air flow on the windward side simply stops causing stalling. In stronger winds this point is reached earlier. This is one of the reasons why sails should be trimmed flatter in windy conditions and deeper in lighter air. The other reason is that side forces also increases with the depth of the sail. Side forces must be reduced as wind speed picks up, until they match the stability of the boat. And the pointing ability upwind gets better with flatter sails.

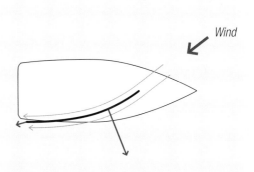

Wind

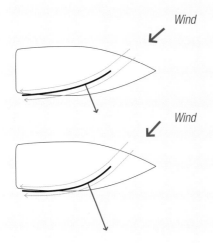

Wind

Wind

Laminar air flow is the term used to describe what happens when the wind follows both sides of the sail profile – from entry to exit without creating turbulence or separation. If a sail is to be efficient, laminar air flow is needed on both sides. This deflects a lot of air and creates a lot of lift force.

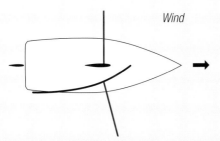

Wind

The keel and rudder produces a certain amount of lift, pulling the boat to windward. The sails produce lift (much more, though), pulling the boat to leeward and forwards.

Drift and weather helm provides an angle to the flow of water. Even though it is small, it is enough for some lift to develop. Too much angle will result in stalling.

SIGNIFICANCE OF THE KEEL AND RUDDER

The keel and rudder actually have cross sections that are designed somewhat like sails or airplane wings and they also attempt to maximise lift and minimise drag. You could say that a sailing boat also has a kind of underwater rig, designed to balance and stabilise the real rig above the water.

The keel does not quite manage to offset lateral forces. There will be an element of leeway. This drift provides an angle of attack between the water and the keel/rudder (except when the boat is sailing downwind), creating lift, much the same way as with the sails. A well designed keel/rudder profile will provide less leeway.

Apart from preventing leeway and creating directional stability, the keel (at least in keelboats) ensures stability. More about that on page 106.

HOW DOES LIFT AND DRAG WORK ON DIFFERENT POINTS OF SAIL?

There is always a calculation behind the progress of the boat. The two components lift and drag will change character according to the wind angle. Here you can see how it works on all three points of sail.

UPWIND

To non-sailors it seems impossible to sail against the wind and it almost is. To make it work, sail profile and angle of attack have to be set within fairly narrow margins. The combined forces of lift and drag have to end up with a positive result. The task is to create a force able to move the boat forward through the water, even with the wind angle almost from ahead, trying to blow everything backwards. This really requires the power of deflection to be optimized and directed precisely. Upwind is by far the most challenging angle to sail, but if you do it right, progress can actually be amazingly effective.

REACH

On a reach it is not as critical if sail profile and angle of attack are not perfectly set. Drag force is no longer directed backwards, but much more sideways (to leeward) and lift force is directed much more ahead. Reaching is the most effective point of sail, because the resulting force of lift and drag is directed in a very favorable direction. In addition, sails can easily maintain a laminar airflow. This combination produces a lot of power!

DOWNWIND

Sailing downwind is not as effective as you might think, even though the wind is coming from behind. The laminar airflow is more or less disrupted – sails are stalled. This happens because the angle of attack is far too big. The wind can not reach the leeward side of the sail and turbulence and separation will also occur on the windward side.

Another factor is that relative wind (explained later in this chapter) on this point of sail drops, as the boat speed increases. Theoretically, downwind sailing is actually the least effective point of sail. It is not all bad however, because the drag forces, which normally act as resistance and lowers the boats speed, is now pushing the boat forward. Moreover, it is common to use large, deep sails (spinnakers or gennakers) downwind, in order to increase the sail area and create more power.

HOW TO READ THE VECTOR DIAGRAMS

The vector diagrams on the next page show how the combined forces of lift and drag affect the boat upwind, reaching and downwind respectively. Boat speed is shown by the thick, black arrow – the length of the arrow indicates the speed. Speed depends on the "resulting force" - the sum of lift and drag.

The resulting force is shown by the dark red arrow. The two red arrows are respectively drag (left) and lift (right). Both lift and drag will change direction and strength, depending on whether the boat is going to windward, reaching or running. The resulting force will be different, both in direction and strength.

UPWIND

Drag forces mainly pull the boat backwards. Lift pulls the boat more sideways than forward. The resulting force is large, but does not have a particularly beneficial direction. Progress is limited and vulnerable – it does not take much to stop working. Effective upwind sailing requires correctly trimmed sails and precise steering.

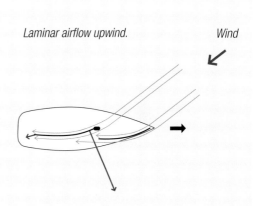

Laminar airflow upwind.   *Wind*

Laminar airflow reaching.

REACHING

Drag forces are now directed more sideways than backwards and lift forces are directed more forwards. The resulting force is large and directed well ahead. The sails still maintain a laminar flow. Reaching is the fastest and most effective point of sail.

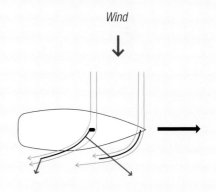

Wind

DOWNWIND

Drag is suddenly a force that helps the boat move forwards. Lift force has an effective direction, but the force is low. Boat speed reduces apparent wind speed (see next page). Laminar airflow is more or less disrupted. Running with the wind is ineffective, but because drag is now contributing positively and the sails are often larger, it still works well.

Downwind, laminar airflow will be partially disrupted.

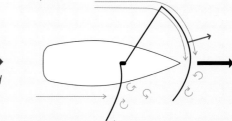

Wind

APPARENT WIND

The wind interacting with the sails is a mixture of two components: the wind that blows when standing still and the wind you create by moving forward.

The wind that works on the sails is called apparent wind. It consists of two components, true wind and induced wind. True wind is the wind that blows across the deck when the boat is stationary. Induced wind is the wind the boat creates by moving forward. Induced wind will of course always come from straight ahead. Apparent wind is a combination of these two. It is the fuel for the sails!

ANGLE AND STRENGTH

Every time that heading or boat speed changes and every time true wind changes, either in strength or direction, the "mixing ratio" between induced wind and true wind will change too. This will affect apparent wind – both strength and direction. To distinguish between the two, we use the terms apparent wind speed and apparent wind angle.

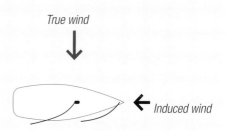

True wind

Induced wind

True wind and induced wind can easily be added, as shown by these two arrows (vectors). Length represents wind speed. Direction represents wind angle.

True wind + Induced wind = Apparent wind

A traditional sailing boat (below) does not move very fast and does not create much induced wind. Apparent wind speed and direction is therefore not so different from true wind speed and direction.
A sports catamaran (left) often is much faster and creates much more induced wind. Apparent wind speed can be much stronger than true wind speed and apparent wind angle can be much smaller than true wind angle.

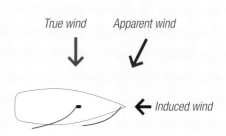

True wind   Apparent wind

Induced wind

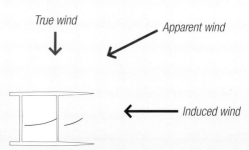

True wind   Apparent wind

Induced wind

APPARENT WIND: THREE HINTS

1 Apparent wind angle is always smaller (directed more forward) than true wind angle. The faster the boat, the bigger the difference.

---

2 If true wind angle is less than 90 degrees (forward of abeam), relative wind speed is always stronger than true wind speed. It seems to be more windy than it really is.

---

3 If true wind angle is more than 90 degrees (abaft the beam), apparent wind speed is always lower than true wind speed. It seems to be less windy than it really is. This only applies to traditional displacement vessels. Planing boats will experience stronger wind across the sails, even with true wind abaft the beam.

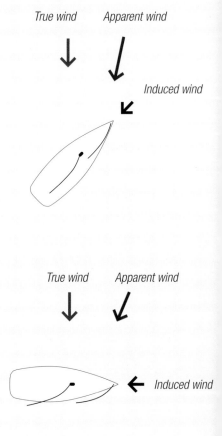

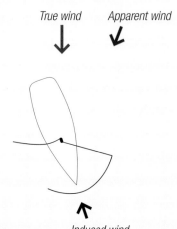

Apparent wind (dark red arrow) is the result of true wind (red arrow) and induced wind (black arrow) combined. The length of the arrows indicates the different wind speeds.

THE SAILS:
MATERIALS AND CONSTRUCTION

Parts of the sail 62
Materials in sails. 64
How are sail profiles made?. 68
Triangular or square mainsail?. 70
Spinnaker and gennaker profile 71

PARTS OF THE SAIL

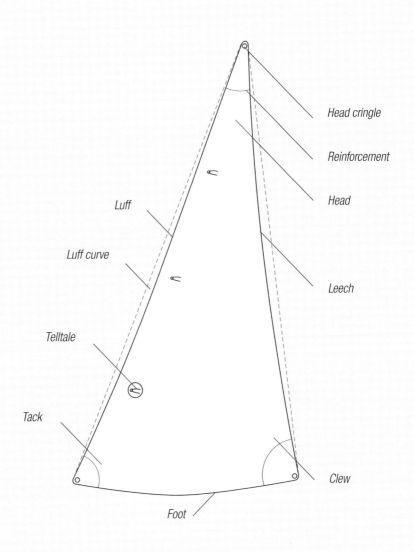

Head cringle

Reinforcement

Head

Luff

Luff curve

Leech

Telltale

Tack

Clew

Foot

HEADSAIL
("Genoa" if overlapping mainsail.
"Jib" if not overlapping mainsail)

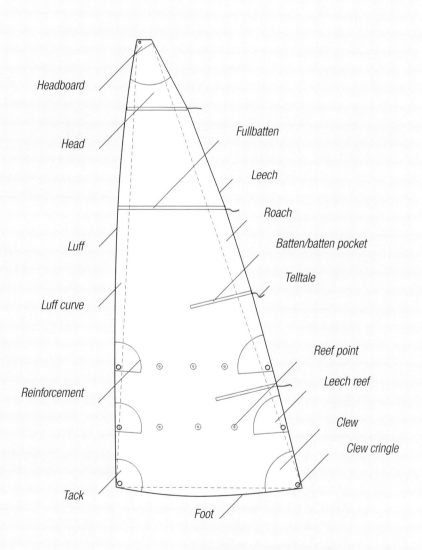

Headboard

Fullbatten

Head

Leech

Roach

Luff

Batten/batten pocket

Telltale

Luff curve

Reef point

Leech reef

Reinforcement

Clew

Clew cringle

Tack

Foot

MAINSAIL

MATERIALS IN SAILS

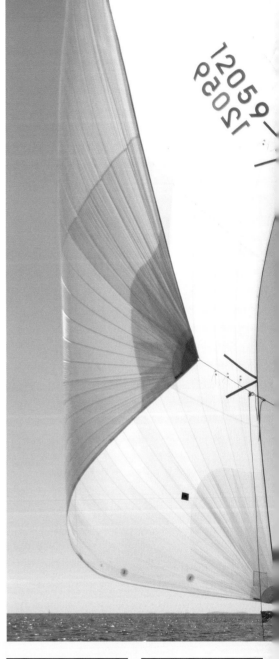

Different sail cloths for different applications

There is a wide range of sail cloth available, suitable for different applications.

When sailmakers choose cloth for a sail, they will typically look at the following parameters:

- Dynamic stretch (flexibility).
- Permanent stretch (propensity to "creep", or extend over time).
- Strength in tensile direction (breaking strength under load).
- UV durability (ability to tolerate sunlight).
- Durability in relation to stress (flapping, packing etc.).
- Weight
- Price

A high score on all these points is not possible. The easiest solution may be of course, to accept a very high price. Other than that, it is a matter of choosing which compromise you prefer.

Cruisers will typically want long lasting and not too expensive sails. As a consequence, sailmakers will not choose cloth with high specifications for stretch, resistance and tensile strength. In other words you will accept reduced performance, especially for the first few seasons, in order to get a longer lasting and more affordable product. Sailors heavily into racing will be more concerned with maximum performance in a new sail. This priority will typically lead to more expensive sails, probably with a shorter lifespan.

Dacron

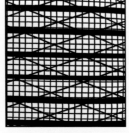

Carbon

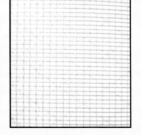

Kevlar

Nylon/spinnaker cloth

The major point when choosing material for a new sail is to find the best possible relation between durability, performance and price. Today it is possible to have laminate sails wiith the same durability as woven dacron/polyester sails, but they cost considerably more. This is why a lot of cruisers with mid-size boats still choose traditional woven sails with a somewhat lower performance.

DACRON OR POLYESTER SAILS

Dacron (originally a product name for polyester sails) is woven with a technique where yarns in the length direction of the cloth (warp) are interlaced the transverse yarns (fill). Longitudinal yarns run over and under the transverse yarns. The main problem with this technique is that when the sail is subjected to load, warp will compress every intersecting point where they cross fill. The result is an extension (stretch) in the longitudinal direction of the fabric. Dacron will stretch even more diagonally, since no yarns are running in that direction. The industry has been working on this issue for many years and there is an ongoing development. Yarns with larger diameter or larger density can be used in the transverse direction of the cloth, or the mesh can be filled with a stabilising resin.

Dacron sails cannot achieve the same form stability as laminate sails fabricated in modern fibers. Dacron is relatively affordable and in a good quality form, it is a very robust material.

LAMINATED SAILS

Laminated sails are extremely stable. They are constructed in a combination of several layers:

"Scrim" is a layer containing the fibers – carbon, kevlar, technora, dyneema, vectran or other high-tech fibers (often in a combination). These fibers are laid out aligned with the load directions in the sail. "Film" is a layer of isotropic plastic material which stabilises the fibers. To make the sails more robust, they are usually made with one or two layers of "taffeta" as well, a woven fabric that provides protection against UV light and other stress influences. This woven fabric layer also contains fibers, normally forming simply a very thin dacron cloth.

The different layers are laminated or glued into to large sections of laminated flat sailcloth. This laminate cloth is then cut into profiled panels and built into a sail using advanced adhesives or traditional seams. A few lofts are able to mould the sail in large machines, all in one piece. The processed sail is finished in a more traditional manner, with reinforcements, reefs, cringles and so on. Laminated sails are typically very costly and their characteristics vary widely. Technological developments have enabled better products and some laminated sails are very durable and well suited for cruising. But generally these sails are performing at their best for a relatively short period of time. Delamination over time has so far been a known weakness for most types of laminate sails.

NYLON SAILS

Nylon is a very different material, only used for spinnakers and gennakers. Unlike other types of sailcloth it is very flexible. Headsails and mainsails need to be as dimensionally stable as possible and are subjected to constant high loads. That is not the case with spinnakers and gennakers. They are flying sails and designed and intended for a certain amount of flexibility. Nylon is also lightweight and has a high tensile and tear strength. A main drawback of nylon is that it is not very UV or moisture resistant.

Nylon cloth is basically woven in the same way as traditional dacron, with transverse yarns interwoven 90 degrees to the longitudinal yarns, again giving the fabric reduced tensile strength across the width and even less diagonally.

HOW ARE SAIL PROFILES MADE?

Sailmakers create profile in mainsails and headsails basically the same way. If you understand how it is done, sail trim becomes a little easier.

Profile in sails is created using two types of curvatures: A curve following the luff and a curve cut into the sides of the panels, before the sail is assembled into one piece.

CURVES IN THE PANELS

Panels can be laid horizontally, vertically, or in a combination of different angles. The most common is the so-called "cross cut", meaning horizontal panels. The panels are cut with one straight edge and one slightly curved. When they are taped and sewed or glued together, a depth or shape will appear in the sail. This profile, resulting from curved panels, is more or less permanent and hard to influence with sail trim. The curves are very small, usually just a few millimeters per meter.

LUFF CURVE

Luff curve (i.e. the second type of curve) is the arch cut in the luff of the sail. In the mainsail, luff curve is "positive", meaning that it "bulges outward" in the middle, as you can see in the illustration. In a headsail, luff curve is usually "negative", meaning that it is cut opposite of the luff curve in the mainsail. The headsail luff curve goes inwards. In both cases the luff curve creates a profile in the sail, but the idea is to adapt the luff to the conditions it will be likely to meet when in use.

The luff curve in the mainsail will follow a mast that bends to a certain extent, primarily longitudinally. The luff curve in the headsail will follow a forestay that is pulled backwards and pushed to leeward to some extent, affected by wind pressure. The forestay will curve and this curve is called "sag". Both sails are designed to adapt to circumstances that will affect them when they are set and subjected to wind pressure.

ADJUSTING THE PROFILE

Luff curve provides a profile that is very easy to change. In the case of the headsail, this is mainly done by adjusting forestay tension. In the mainsail, it is mainly about increasing or decreasing mast bend. In addition, the forward part of sail is sensitive to luff tension, i.e. how hard the halyard is set.

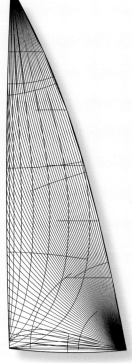

In laminated sails, sail designers try to make sure fibers are oriented in directions aligned with the stress or load paths in the sail. They also make sure there are more fibers in high load areas.

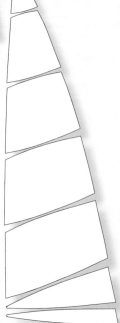

Panels are curved in front – this becomes the luff curve and will be cut when the panels are assembled. Panels are also curved in the other direction, here horizontally (cross cut). These two different curves create the profile in the sail.

TENSION IN LUFF, LEECH AND FOOT

Higher luff tension will move the deepest point in the sail forward and at the same time flatten the sail and open the leech. More leech tension will decrease twist (twist is the tendency of the sail to open the leech increasingly towards the top, see more on page 108). More foot tension will reduce profile and flatten the bottom section of the sail.

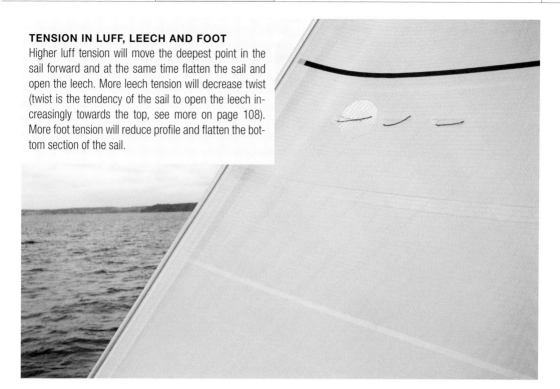

MAST BEND
AND FORESTAY TENSION

Luff curve should fit "rig curve". The headsail is designed to fit a certain forestay sag and mainsail is designed to fit a certain mast bend. By changing rig trim, you will alter the sails profile. Greater forestay tension will pull out (reduce) the profile that luff curve creates in the headsail. Greater mast bend will pull out (reduce) the profile that luff curve creates in the mainsail. This will be discussed in more detail later, but here is a very short explanation:

- More mast bend flattens and opens the mainsail.

- More forestay tension flattens and opens the headsail.

These effects cooperate well. As we shall see, the backstay is a key tool used to regulate both.

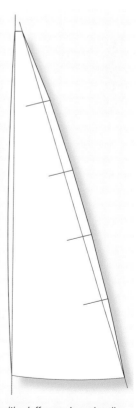

Negative luff curve in headsails   *Positive luff curve in mainsails.*

TRIANGULAR OR SQUARE MAINSAIL?

Traditional mainsails are triangular. Most of them are designed with a certain positive, arched curve in the leech, supported by battens (roach). In some mainsails this is taken even further. These are designed with an extra large surface at the top, often referred to as a "fathead".

A lot of modern, fast sailing yachts (especially multihulls) have mainsails with very wide heads. Some are elliptical, others almost square. This type of mainsail, often called fathead or square top, offers a larger area at the top and are more powerful for a given mast height, especially downwind.

A LOT OF TWIST

Theoretically these sails can also minimise "vortex", a swirling airflow formed at the top of the mainsail, creating drag (see page 110).

The challenge with these sails is to prevent them from twisting too much. The more sail area that ends up outside of a straight line between the rear corner and the top of the sail, the more twist you are likely to have. This also goes the other way: For the same reason, furling mainsails without battens have almost no twist at all. They are cut with a negative leech curve or no roach at all.

MORE POWER DOWNWIND

It should be obvious why a mainsail with a big head provides more power downwind. The wind speed is higher in the top of the sail and more sail area here provides more power. But if these mainsails are built and trimmed properly, they can actually be very efficient upwind too, despite the fact that we typically don't want too much pressure in the top of the rig when the wind angle is tight. Side forces and heeling will quickly become a problem.

HIGH SHEET TENSION

But fathead mainsails twist very easily. As a consequence, they quickly get rid of any excess power. The problem may actually be that they twist too much. The big head will then induce more drag than lift. When that happens, you are in fact, dragging a lot of passive sailcloth through the air. Only the lower, front part of the mainsail produces real lift. If they are to be effective upwind, fathead mainsails require a lot of sheet tension along with rigid, well tensioned battens.

All things considered, fathead mainsails will work best on planing boats, where the crew are continuously trimming them upwind. They also work well on cruising catamarans, because of the boats greater stability.

PROBLEMS WITH THE BACKSTAY

These sails are complicated to build, if they are to work well. They also make greater demands on the rig. Mast and standing rigging must be dimensioned to handle the larger forces at work at the top of the mast. There are also practical problems. The sail can not pass the backstay, at least not with backstay tension on. On some boats, the backstay is connected to a flexible "arm" (a batten or similar) mounted on the top of the mast. When the backstay is released, the arm lifts the backstay, allowing the fathead to pass. Double backstays are another option. Some boats even have no backstay at all, but use running backstays or simply rig tension from capshrouds and swept spreaders to keep sufficient longitudinal rig tension.

TRIANGULAR EASIER TO MANAGE

A traditional triangular mainsail is easier to build and therefore cheaper. It also has the advantage of being able to pass easily underneath the backstay in tacks and gybes. Twist in these "normal" mainsails is easier to control. In fact, it can often be a challenge to make them twist enough. More on this on page 108.

SPINNAKER AND GENNAKER PROFILE

When it comes to spinnakers and gennakers, not only sail cloth properties but also the functioning of the sails is very different from headsails and mainsails. But why are these sails shaped the way they are?

Sails made of nylon, in other words spinnakers and gennakers, are an entirely different species than sails made to go upwind. Spinnakers and gennakers are all about being light, large, deep and flexible. They are invariably set flying and are rarely exposed to big loads. They do not have to be adapted to a mast curve or forestay and are mostly used on relatively open angles where they do not need to be trimmed particularly flat. For these reasons they can be designed with large curves, both in panels and luff/leech/foot. Low weight is essential and flexibility is high, so they don't burst as easily when they collapse and fill in gusts.

OTHER FABRIC FOR SHARPER ANGLES

You can also build spinnakers and gennakers in modern fibers, but this has not become as popular as in mainsails and headsails. Cuben fiber is often used on ambitious racers with high budgets. Flat gennakers designed for very sharp angles (eg. Code Zeros) will typically benefit from a different material than nylon. The need for high tensile strength returns when the sail is used at sharp wind angles and trimmed with high luff tension. Code Zero's tend to be produced in thin and light laminate construction, although nylon in various configurations still dominates the market

for more traditional spinnakers and gennakers. There are actually two significant advantages with cheaper, more flexible spinnaker cloth. It fills more easily in a light breeze and tolerates a lot of punishment in strong winds.

ENTRANCE AND EXIT

Although spinnakers and gennakers are normally built in the same material and have somewhat overlapping functions, there is one fundamental difference: The gennaker is in reality like any other sail, that is, an asymmetrical triangular sail where the three corners don't change places. It is a flying sail, just like the spinnaker, but both sheets are attached to the same corner, like in a headsail. The gennaker is designed with a defined entrance and exit, creating a difference between the luff and the leech. Maximum depth is slightly ahead of the center (approximately 40–45 %).

CAN SAIL BOTH WAYS

This is not the case with a traditional symmetrical spinnaker. Here the luff and leech change every time you gybe. A spinnaker can be sailed both ways. It is therefore symmetrical: The deepest point is in the middle (50 %) and both sides are cut identically.

THE MOST COMMON SPINNAKER DESIGNS

CROSS CUT
Panels are laid horizontally (cross cut) but this causes a lot of uneven stretch at the head of the sail, where forces work diagonally. It works well with low wind angles (running) but in heavy weather or on a reach, a cross cut spinnaker will easily deform.

RADIAL HEAD
A diagonal pattern radiating from the head (radial head) makes the sail more stable in strong winds and when reaching. If the sail is specially designed for sharp wind angles even this design can run into problems. Reaching spinnakers need even better form stability..

TRI RADIAL
Tri radial spinnakers can better cope with diagonal loads across the sail, since the panels are laid out in the three dominant stress directions. These sails work well on all points of sail, even reaching. However, the more laborious production process involved makes them more expensive.

06

SAIL HANDLING:
HOIST, TAKE DOWN, FURLING AND REEFING

Hoist, take down, furling and reefing 73
Mainsail 76
Headsail 78
Furling systems 80
Furling main 83
Furler or snuffer? 84
Reefing 87
Reefing the headsail 88
Reefing the main 90
Reefing the mainsail downwind 94
Sail maintenance and repair 96

SAIL HANDLING:
HOIST, TAKE DOWN, FURLING AND REEFING

Good routines and a good set up for handling sails helps reduce wear prevents damage and helps avoid potentially dangerous situations. It also makes sailing more comfortable, enjoyable and efficient.

There is usually more wear and tear on the sails when they are hoisted, taken down, furled or reefed, than when actually sailing. It is also in these situations that many stressful and potentially dangerous situations on board occur, especially when the wind is up. It makes sense to have some well considered procedures and a rig that is set up to facilitate the handling of sails as easy and efficiently as possible.

FURLING SYSTEMS ARE POPULAR
Furling systems are becoming more and more popular, not only for headsails, but for mainsails and flying sails as well. These systems have great benefits and a few disadvantages that it's useful to know about. For example the limitations of a reefing system with furling.

In this chapter we will look at hoisting and taking down the different sails. We will also look at furling and reefing with furling systems, as well as traditional methods of reefing.

HOISTING AND TAKING DOWN FLYING SAILS
Handling flying sails (gennakers and spinnakers) can be a complex process, especially so with spinnakers. The subject is fairly broad and falls a little outside of the scope of this book, but the different methods are well described in other books or online. In this chapter we have chosen to limit the subject to handling of spinnaker and gennaker using furlers or snuffers/socks.

MAINSAIL

Sliding cars and lazy jacks! This is how your mainsail runs easily up and down.

Cruising boats are often sailed by a limited crew – often just twopeople. A mainsail with luff rope fed into the mast track be-comesdifficult to control in strong winds. Because there is nothing holdingthe free part of the sail, it can quickly end up all over the deck.Sliding cars will make sure the mainsail is under control at all times,not only during hoisting or dousing, but also when reefing.

AGAINST THE WIND
As you probably know, the mainsail has to be hoisted with the bow more or less directly into the wind. Otherwise it gets jammed against the mast, shrouds and spreaders. Most people use the engine to keep the nose up against the wind. Go as slowly as possible. Some will even reverse, to reduce flapping and wear on the sail.

If you already have a headsail up and want to set the main, that's fine too. Sail close-hauled upwind and make sure you have enough speed to retain steering ability.

If using a winch, stop if you detect an increase in load and determine the cause. A good idea is to have one person pulling on the halyard at the mast and another taking up the slack at the winch.

CONTROLLING THE BOOM
When the sail is to come down, head into the wind again. Pull in on the sheet – enough to keep the boom from swinging from side to side, but not so much that the sail fills. Release the halyard, let the sail come down and pack it on the boom immediately. It is a good idea to start from the stern, using relatively large folds draped on both sides of the boom. The sail ties should be tight enough to hold the folds in place, but not so tight that they cut into the sail.

Left: Lazyjacks will keep the mainsail under control when taking it down.

Middle: Sliding cars are a big help when setting or dousing the mainsail. They keep the front part of the sail under control.

Right: Head more or less into the wind, when the mainsail goes up or down.

TIPS:

- Lazy jacks, possibly with a lazy pack (see photo, bottom left) are very useful tools.

- It is a good idea to have lazy jacks arranged to be pulled forward to the mast when not in use. Otherwise they will wear on the leeward side of the sail while sailing.

- Lazy jacks works best on full battened mainsails. Mainsails with short battens will have a tendency to hook the lazy jacks when the leech flaps around during take down.

- If the outhaul is trimmed hard it should be loosened when the mainsail is back down. Take the halyard off and hook it somewhere at the back of the boat (for instance at the end of the boom). This prevents it from slamming the mast and wearing unnecessarily.

- Most often, the worst damage done to a sail is UV rays. Always use a sail cover to protect against the sun.

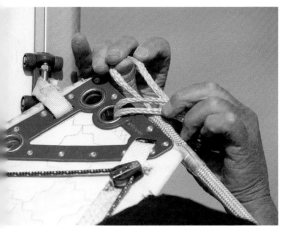

HEADSAIL

Furling headsails are easy to use and reliable – provided the furling system is handled properly.

Furling headsails are standard equipment on most sailing boats. Racing boats and some smaller vessels use headfoils or stay hanks, primarily to be able to change headsails faster. A furling system makes it very easy to handle the headsail. You release the furling line while pulling in on the sheet. When you want to furl the sail back in, you do the same thing, just in the opposite order.

Nevertheless, there are a few tricks that might be useful to make sure the furling system does not get stuck or is subjected to unnecessarily high loads.

NICE AND EASY

Feed out the furling line slowly. Add some resistance, make sure the sail unfolds at a steady and slow pace. If it unfurls very quickly you will increase the risk of overriding turns on the drum and the furler may get stuck when you furl the sail back in. Jerks and sharp impacts may overload and damage the system. Sheet in once the sail is out, to minimise flapping.

DEPOWER THE SAIL

Depower the headsail before furling. Minimising pressure makes it much easier to furl and the chance of damaging the furling gear is greatly reduced. In situations where there is no power at all in the headsail (very light winds), it is necessary to keep a little tension on the sheet, to help the sail furl fairly tight and to make sure the furling line furls evenly on the drum with more or less even tension. This will prevent the system from fouling when the sail is unfurled again. If you try to furl the sail back in while sailing to windward in heavy weather, you will put great strain on the furling gear and sail. A solution might be to head down on a broad reach, low enough for the mainsail to cover the headsail. Now the headsail will be tamed and easier to furl.

WHAT IF IT GOES WRONG?

What if it the system malfunctions when the sail is completely or partially unfurled? Basically there are three options: You can identify and fix the problem. You can take down the sail by releasing the halyard and bring it down through the luff profile (not possible if the sail is reefed or partially unfurled). Or you can turn the boat around itself repeatedly until the sail is rolled back on the forestay (requires removing or releasing the sheets). In extreme situa-

Tips: A good habit is to look up the mast when sails are set and make sure halyards run as they should all the way up. They can easily get tangled.

tions, for instance if the headsail is ripped out of the furler during a storm, it may be necessary to drop or cut the sheets and let the wind shred the sail to pieces in the wind.

UV PROTECTION
Sacrificial UV-strips on the leech and foot are important. Most headsails are stored rolled up on the forestay. Without UV protection the sun will destroy the exposed parts of the sail in the course of a few seasons. A headsail sleeve is a good supplement, but if it is to have any effect, you have to actually spend the extra minutes it takes to put it on at the end of the day – every time.

JIB HANKS
Jib hanks on the forestay are a simple, inexpensive and robust system used for centuries. The original piston hank requires two hands and if you feel you need a free hand on the foredeck, it may be worth considering the newer models, that can be operated with one hand. Whichever you choose, using hanks means work on deck. You have to take down the old sail and remove it completely before a new sail can be hoisted. This is not always an easy job.

HEADFOIL
A headfoil is similar to a furling system, only without the furling drum. The headsail luff is fed into one of the two tracks in the foil and hoisted, much the same way as the mainsail. Usually, there are two halyards and two tracks and it is best to use the corresponding tracks and halyards

(on the same side). If you use the opposite halyard you may create an unwanted diagonal tension between the sheave and the luff track when the sail is fully hoisted. It could even be difficult to bring the sail back down again.

A significant advantage for racing boats is that you can set a new headsail before taking the old one down. This way there is very little loss of speed when changing sails. Going upwind, the easiest way is to hoist the new sail inside of the old, on the windward side. Attach the lazy sheet (leeward sheet) to the new sail and tack. Now the old sail is on the windward side and can easily be pulled down inside of the new.

OUTSIDE OR INSIDE?
When racing in a straight line, everything has to be done on the same tack. That means hoisting the new sail in whichever free slot there is and then take down the old sail – outside or inside. Remember to use a pre-feeder (see picture) to make sure the luff tape does not tear while hoisting. The foredecker needs to communicate with the crew on the halyard.

On a reach or in severe conditions, it may be necessary to take down the old sail before the new one goes up. The new sail should at least be made ready on deck, so the halyard can be moved straight over, minimizing the time without a headsail.

Tips: Prefeeder helps feed the luff into the headfoil and prevents the luff tape being torn when the sail is hoisted.

FURLING SYSTEMS

It is very comfortable to be able to furl the sails instead of hoisting and dousing them. But there are some inherent problems with furling systems and knowing them could be useful.

All sails can be furled. A lot of different systems have been designed and because of the constant development most sails can be rolled with a reasonable degree of reliability. Furling headsails have been standard equipment for many years and mainsails have been furled on booms for even longer – actually more than a hundred years. The furling mainsail is a more recent invention. Furling mains have become increasingly popular as the craving for more and more comfort has grown. Besides, people generally sail in larger boats with smaller crews than before and this makes easy sail handling more important. Gennakers and spinnakers can also be furled, even though a snuffer will often turn out to be a cheaper and more reliable solution.

POORER SAIL TRIM

No matter how good they are, furling systems do have their drawbacks. This especially applies to gennakers and spinnakers and to some extent mainsails. The financial cost can be substantial and for mainsails, sailing performance is to some degree compromised. You also introduce a rather complicated mechanism that may stop working. If this happens, it will probably happen in situations where loads are big, conditions are demanding and the last thing you want is a technical problem.

GREATER COMPLEXITY

In other words, the disadvantages, are a boat that is not performing quite as well as it otherwise would have done and where the risk of technical problems is somewhat increased. Still, most cruisers put greater emphasis on easy handling than sailing performance. And the risk of furling systems breaking down or causing problems actually seems to become smaller as systems are improved year by year.

TIPS:
You can improve the profile in a furling headsail when reefing. One crew pulls the foot of the sail down while another operates the furling line. This prevents the foot from creeping up on the furl, which would otherwise give the sail more fullness than you want from a reefed sail. In other words: It helps flatten the sail.

TIPS:
On a reach it is easy and very efficient to adapt the sail area to wind conditions by furling the headsail partially. Just remember to move the genoa lead car forward, to adjust the sheeting angle.

Furling mainsails with vertical battens (top) make it possible to build mainsails with a roach. They perform better, but battens can get stuck in the mast track.

Furling mainsails without battens (bottom) have to be cut with a negative leech. Sail area is reduced and they do not twist as well as mainsails with a roach.

FURLING MAIN

There are mainsail furling systems both for masts and booms. Both solutions provide easy handling, but sailing performance will suffer to some degree. Proper handling will reduce the risk of technical problems. Here's how.

Furling mainsails work best when fully unfurled, just like furling headsails. Reefing systems for furling booms have especially been known for one weakness: Tension in the foot is lost as the sail is furled. This makes the sail fuller, especially the bottom section. The leech will also be more closed – the opposite of what you want from a reefed sail. Some new furling boom systems are built with conical cylinders that work better.

Still, traditional reefing with reefing lines is a solution that provides a more efficient result. The sail simply has a better profile after reefing. The method is described later in this chapter.

FLATTER SAIL AND SMALLER SAIL AREA
Furling masts are specially constructed and their extra weight will slightly affect boat stability. They also need specially built mainsails. The biggest problem with these mainsails is that horizontal battens can't be used. This makes it difficult to design a good mainsail. Roach (extra area outside a straight line from clew to head) gives a larger sail area, more power and better twisting abilities. But a roach can not work without battens. On the contrary, without battens the leech has to be cut with an inward, negative curve and the result is a smaller mainsail without the same ability to twist. The sail also has to be fairly flat, to be rolled into the mast. The mast curve can not be trimmed to the same degree by bending the mast. A larger diameter mast combined with a shorter sail chord (distance from luff to leech) also means significantly less laminar airflow and lift production.

VERTICAL BATTENS
Sail designers have come up with furling mainsails with vertical batten – some even fullbatten sails, making it possible to construct sails with some roach. These sails perform better, but the battens can create new problems. Battens can get stuck in the mast track and there is a lot of wear on batten pockets when the sail is furled and unfurled.

Laminated mainsails are better suited for furling systems than dacron sails. A more stable mainsail minimises problems when furling.

PROPER HANDLING
Despite the limitations, a lot of cruisers are very happy with furling mainsails as the work and trouble of hoisting, reefing, taking down and packing the sail are gone. Many technical problems are caused by improper handling. The outhaul must be tensioned a little as the sail is furled, especially in light weather and the boat should head approximately 60 degrees to the wind (somewhere between a close reach and close hauled). It helps always to furl with the wind in a direction that brings the sail through the mast track on the windward side, where there is more space. One side is inevitably better than the other, due to the fact that the sail is always furled in the same direction inside the mast.

If you find sailing performance important and sail trim interesting, a furling mainsail is a bad idea. They provide comfort and easy handling, but not without compromising mainsail efficiency.

Cross section og mast with furling mainsail

Right: The sail is furled in the correct direction – without causing more friction than necessary.

Wrong: The sail is pulled into the mast around the edge of the mast track. This increases friction and wear on the sail.

GENNAKER AND SPINNAKER

FURLER OR SNUFFER?

NOR
15039

SNUFFER/SLEEVE/SOCK

Both spinnakers and gennakers can be hoisted and doused using a sock, also called snuffer or sleeve (see photos on top of the page). The sail is stored in the sock and hoisted with the sock still on. When fully up, you "peel" the sock off the sail from the bottom up, using the integrated endless line while adjusting the guy/tackline and sheets. The sock ends up lying compressed on top of the sail.

When taking down, allow the spinnaker or gennaker to collapse in the wind shadow from the mainsail and pull the sock down over the sail, again using the integrated endless line. Now the sail is easier to lower on deck.

A Code Zero is a flatter sail designed for high luff tension and the sail cloth is normally more stable. That makes the sail better suited for a furler system than a nylon gennaker or spinnaker.

FURLERS FOR GENNAKER AND SPINNAKER

These sails are usually difficult to furl with a reasonable reliability.

There are furling systems specifically designed for gennakers and the development of these systems is ongoing. It has been difficult to make it to work well, because the necessary tension in the luff is rarely present. Nylon cloth is very flexible too and the sail has no stay to furl around. A Code Zero – a very flat gennaker for sharp wind angles, can be furled with better results, since the luff is tighter and the fabric more stable.

"Top down furler" is a variation used for both spinnakers and gennakers. Here, the sail is furled from the top down, around a particularly torsionally stable rope attached to the top of the sail. The system can be retrofitted and used with an existing gennaker or spinnaker and has no negative consequences for the function of the sail once it is set. This type of furler has had problems due to stretch in the sail. Furling systems for spinnakers and gennakers will probably need some further development to achieve the necessary reliability. A sock will most likely be a better (and cheaper) solution for most purposes.

Less sail area, lower down. This is how you adapt the pressure in the sails to suit the stability of the boat. More about this on page 106.

REEFING

Sail area must be adapted to conditions. This is high on the list when it comes to seamanship. A reef properly set at the right time makes the boat more comfortable and less stressed. It will probably even go faster!

Most sailing boats are designed and rigged for a wind speed of 12 knots. This is the average wind speed used by boat designers when they try to work out the best combination between stability and sail area. Less wind makes the boat more or less underpowered, more wind makes it more or less overpowered.

To be fair, the spectrum where the boat sails fairly well extends somewhat either side of 12 knots - but there is definitely a need to adapt the forces in the sails to the wind conditions as they change. Much can be done with sail trim, but it has limits.

MULTIPLY BY FOUR!
When the wind speed doubles, wind pressure is not just doubled. The forces working on the sails are four times stronger, because the relationship between wind speed and force is squared. This means that if you start off on a wonderful summer's day in 8 knots of wind and the wind speed increases to 16 knots, the wind pressure in the sails is four times higher. If wind speeds reaches 32 knots, you can multiply the initial force by 16.

NOT ENOUGH SAIL?
The same goes of course when the wind speed drops. You will very quickly lose power and feel the boat sailing considerably slower, simply because there is not enough sail area. In light winds, you can choose to set large spinnakers or gennakers on open wind angles, but hard on the wind there is not much to do. Older boats can be rigged with large overlapping genoas (up to 150%) and they can pull quite well close hauled in light air. But newer models typically have a weak spot here, with smaller, more high aspect headsails.

On the other hand, a little lack of boat speed in light winds might not be a big problem. You could take the opportunity to relax and enjoy life – or start the engine if you are in a hurry.

PROPER TRIM WILL HELP
When the wind builds to considerably more than 12 knots, it is essential to adapt sail area to the conditions. Trimming the sails correctly will bring you a long way and expand the wind range where the boat sails well with full sails. More on this in chapter 8.

Still, in most boats the crew will want to consider reefing when the wind speed reaches 15-20 knots.

A boat with too much sail up is not a comfortable place to be. The risk of damage to equipment and potentially dangerous situations is bigger. Balance and steering capability may suffer. Off the wind it can be very entertaining, if the boat is designed and rigged for high speed and the crew is competent enough to handle the conditions. But upwind it is not only an unnecessary strain on boat and crew, it is also not very effective. An overpowered boat would in most cases have sailed faster – and be more comfortable – with a smaller sail area.

REEF DOWNWIND AS WELL
Traditional displacement boats should also take care to reef on open wind angles. Too much power increases the risk of the boat digging the bow into a wave, or broaching and losing control (see more about this on page 187). Besides, the boat will rarely break loose from its theoretical hull speed, so there is not much speed to be gained.

When sailing downwind it is not so easy to notice an increase in wind speed and you can get caught in a situation where it becomes difficult to go upwind with full sails in order to reef. In fact, it is possible to put in a reef going downwind. This is explained at the end of this chapter.

All in all, it, sailing with too much sail up is not good seamanship.

REEFING THE HEADSAIL

With a furling headsail, all you have to do is ease the sheet and pull on the furling line. But the result upwind is often far from perfect. With jib hanks or a headfoil you have to change to a smaller sail. That is a lot of work, but the result will probably be better.

Reefing a furling headsail is very simple. You ease out the sheet and pull on the furling line until you are satisfied with the resulting sail area. The sheeting point must also be moved forward, to maintain the correct sheeting angle. This is important and may actually be the most tricky part of the operation (depending on how the car on the genoa track works).

LOWER LUFF TENSION
Furler reefing works fine on open wind angles, but is not so good upwind. The sail will usually become fuller and lose the designed profile and upwind there is a very big difference between a sail that is trimmed to the correct profile and a sail that is not. The main problem is that luff tension is lost when the sail is furled. A slightly overlapping genoa, such as 106%, can be rolled in a few turns without deforming too much, but a big, broad genoa with many rolled turns has problems keeping a good, flat profile upwind.

CONICAL FOAM LUFF PROFILE
Many furling headsails are built with a conical foam profile in the luff. This helps pulling some fullness out of the mid-section of the sail when furling. Vertical battens are also quite common. They help in maintaining profile and make it possible to add some extra area in the leech. But the battens are hard on the sail, partly because they slam against the mast and partly because of the wear and tear when the sail is furled. It is rarely furled 100% parallel with the battens. Too ceep acceptable shape, furl only about 1/5 of the foot of the sail

SAILS TO FIT CONDITIONS
If you want efficient sails upwind in heavy weather, the best thing is a smaller headsail, able to maintain a flat and relatively open profile even when sheeted in hard. This sail can be set on an inner forestay, or even flying. When the weather gets rough, you simply furl the big headsail and hoist the smaller one. If you have a rig where this is not an option, you can of course still go upwind with a reefed furling headsail. How well it works depends on how the sail is designed and how old it is. But it will rarely perform as well as a smaller headsail made for heavy weather.

REQUIRES SEVERAL SAILS
Boats with traditional hank-on sails or headfoil are well set up to adapt headsail size to changing conditions. But the process is quite labour intensive and you need several sails in different sizes. To reef, you need to take down the existing, bigger headsail and stow it below deck, before a smaller one can be hoisted. This operation should be carried out before the wind and sea state becomes too harsh. The result, once it is done, will most likely be better than a partly furled headsail.

CHANGING TO A SMALLER HEADSAIL WITH HEADFOIL

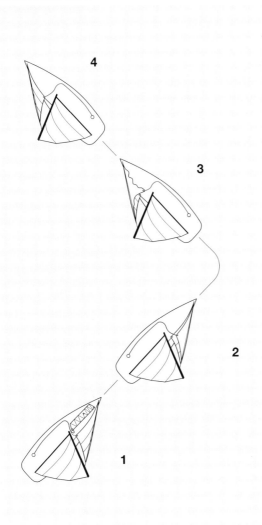

When changing headsails on a headfoil, it is best to hoist the new sail inside the old. Then you can tack – and take down the old sail inside the new while on the other tack.

1. New sail ready on windward side – windward sheet in the new sail.

2. New sail is hoisted in headfoil track on the windward side.

3. Boat tacks. New sail is sheeted on leeward side.

4. Old sail is taken down on windward side.

REEFING WITH FURLER

Full sail area

A little less sail area

Very small sail area. Remember that there needs to be a sheeting point in the deck to fit the sail size.

This is what happens if the sheeting point is not adjusted.

REEFING THE MAIN

FURLING MAIN

Furling mainsails can be reefed without much trouble. See more about reefing a furling mainsail on page 83.

TRADITIONAL MAINSAIL SYSTEM:

There are a few variations of "slab reefing", which is the basic, traditional way to reef a mainsail. No matter how the lines run, the basic idea is to lower the mainsail with the halyard, until the desired reef point in the sail reaches the boom. Then the reef is secured in two places – at either end of the boom.

This system is simple, inexpensive and extremely reliable and gives a better result than reefing with furling systems. Done properly, the sail not only gets smaller in area, but also flatter and thereby even better adapted to the conditions. Over the next pages, the operation is described in five steps, here shown using a reefing hook.

REEFING THE MAINSAIL

1   Sail close-hauled, let out the mainsheet until the mainsail no longer is loaded. The kick should be completely loose. Take care to keep speed enough for steering, just using the headsail.

2   Lower the halyard until the forward reefing cringle can reach the reefing hook at the gooseneck. Go forward (use harness if circumstances require) and put the cringle on the hook. Tighten the halyard back up again. Tighten hard!

3   Tighten the reefing line until the reefing cringle in the leech is close to the boom and the foot of the mainsail is pulled tight along the boom. This requires a lot of tension in the reefing line. Remember to keep an eye on the reefing line and make sure it doesn't pinch or damage the loose part of the sail.

4   Tighten the mainsheet and keep sailing.

SECURING THE REEFED MAINSAIL

A lot of mainsails are equipped with a series of cringles along a line between the reefing points. Here, you can run sail ties or lines to secure the loose part of the mainsail when reefed. They are seldom necessary to use, but if you use them, don't tighten them hard. You run the risk of damaging the sail if the load ends up here instead of on the reef hook and reefing line. There is also the risk of forgetting to take them off when shaking out the reef, which can result in severe damage of the sail. Use these cringles only when necessary, set them loose and take them off as the first thing when the reef goes out.

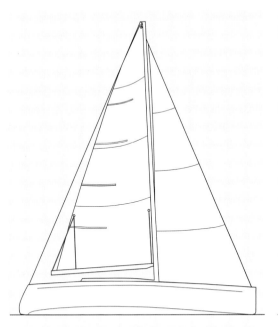

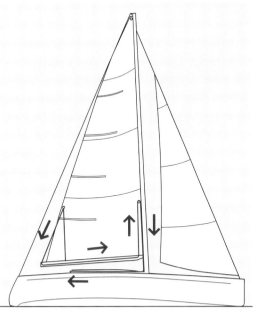

DOUBLE REEFING LINES

A variation where the reefing hook at the gooseneck is replaced with a line led back to the cockpit. This eliminates the need to go on deck to reef. Procedure is otherwise exactly the same.

SINGLE LINE REEFING

Both reefing points are controlled with a single line. The line runs through the reefing point in the leech, then through the reefing point in the luff and then back to the cockpit. Very simple and easy to handle, but with the disadvantage of added friction in the system.

TRADITIONAL MAINSAIL REEFING

Ease the halyard and pull the luff down. Hook the luff cringle in the reef hook or tighten the luff reefing line. Tighten the halyard really well. Now it is time to tighten the reefing line on the boom. That's it – sheet back in and sail on. Here you can see how the mainsail is reefed – shown in five steps.

Ease the sheet enough to take the pressure off the sail. Release the halyard. Let go enough to allow the reefing cringle to be pulled down to the boom.

The cringle is put on the reefing hook. A line through the reefing cringle can be used instead.

When the reef is secured at the mast, tighten the halyard again. Tighten hard! To flatten the sail sufficiently, you will need much more halyard tension than in lighter air. If it is set, ease the kick.

Tighten the reefing line on the boom. The line and the block on the boom has to sit in a position that pulls the foot of the sail flat and at the same time pulls the leech down close to the boom. You may need to ease the sheet a little more to make it possible.

The reef is ready. Sheet in and trim according to the heading.

REEFING THE MAINSAIL DOWNWIND

Not many people are aware of it and very few use this technique, but it is actually possible to reef the mainsail while sailing downwind. It might even be a very good idea.

There are situations where you will be thankful for not having to head into the wind to put in a reef. It could be in high wind and heavy seas, or perhaps on a downwind leg in a regatta. Or if you have a spinnaker or gennaker up and need to reduce mainsail area.

FULL SPEED WHILE REEFING

The good news is that it is actually possible to reef the mainsail without heading into the wind. The procedure might be vary slightly on different boats, but the main thing is to keep some tension in the leech with the reefing line and do it step by step, little by little. It may take some time, but that doesn't matter as the boat will keep sailing while the reefing takes place. If you have swept spreaders, it may be necessary sheet in a little, to ease the mainsail off the spreaders.

ALTERNATE BETWEEN HALYARD AND REEFING LINE

Release the halyard a little – but only a little! The main will normally come down by itself, even when there is pressure in the sail. If it doesn't, give it a hand. Now tighten the reefing line until the leech is fairly tight again. Let go on the halyard once more and carry on alternating between the reefing line and the halyard until the mainsail is lowered to the reefing point you need. Tighten the halyard again (with some caution, there will be pressure in the sail and resistance in the mast track).

That's it. You have reefed the mainsail, sailing at full speed downwind.

← A preventer stabilises the boom and will give you a second chance to prevent an involuntary gybe. The preventer is normally led all the way to the bow, but it can also be led further back, like this – especially if the purpose is to stabilise the boom on a reach.

PREVENTER

A preventer is a strong line running from the end of the boom to the bow. The best thing is to have it run through a block on the bow and lead it back to the cockpit, so it can be adjusted from there.

The preventer's job is to stabilise the boom when sailing downwind. In a swell the boom can easily start swinging backwards, disturbing the mainsail. On a dead run, the preventer works as a kind of safety valve for involuntary gybes. If the mainsail starts to backwind, the preventer will

stop the boom, giving the helmsman a second chance to steer back on course.

If strong wind really gets hold of the back of the mainsail, the preventer can't prevent a difficult situation. The mainsail can be pressed hard against the centerline of the boat and the boat loses speed. In this situation you will probably lose steering ability and be forced into an involuntary gybe. The boat is rotated beam to the wind and the pressure on the preventer will increase until something breaks. If this happens, it is important to release the preventer as soon as possible, to let the boom over on the opposite side.

There are products on the market made to slow the boom down during gybes – so called boom breaks. A line running across the boat is pulled through a type of friction lock on the boom. This can be a nice supplement to the preventer.

SAIL MAINTENANCE AND REPAIR

Sails are expensive and have a relatively limited lifespan. This makes them an investment it really pays to take care of. Maintenance, regular checks and minor repairs can extend the life of your sails considerably.

+ LIMIT FLAPPING

The lifespan of sails is reduced minute by minute when they are allowed to flap. One of the best things you can do for your sails is to prevent this from happening more than absolutely necessary. Avoid going fast under engine upwind when setting the mainsail. In fact, it is a good idea to reverse instead, to reduce windspeed over deck. Fullbatten mainsails will be less prone to flapping and last longer. Sheet the headsail quickly after unfurling or tacking.

Make sure leech and luff lines are tightened enough to prevent vibration and remember to ease them off again before the sail is furled or packed. Laminated sails are particularly sensitive to flapping.

+ PROTECT FROM THE SUN

UV rays constantly break sailcloth down when exposed. The sun is responsible for destruction of more sails than any other single factor. Use a sail cover for the mainsail and

sacrificial UV strips in leech and foot for furling headsails. Do not leave sails in the sun longer than you need to.

+ RESPECT WIND SPECTER

Mainsails and furling headsails are mostly designed for wider ranges of weather. This is not the case for headsails with jib hanks or luff tape. They will be built for a specific wind range. The same applies to spinnakers and gennakers. Do not use these sails in stronger winds than they

are built for! Once may be enough to permanently deform or damage the sail. Make a note of the wind range on the clew, visible to the crew. This information can be obtained from the sailmaker.

+ PROTECT AGAINST WEAR AND SHARP EDGES

Make sure the sails are not damaged by wear and tear over time, or by sharp edges such as split pins on rig bolts or turnbuckles, spreaders, shrouds, stanchions or the like. Lazy jacks or running backstays and check stays will chafe the main if left hanging on the leeward side over time. Also

be careful when the sail is moved around or stowed below deck. You can also prevent wear or damage to the sails by ensuring that all sharp objects on the boat and rig are covered with tape or other types of protection.

+ PROTECT FROM SALT AND MOISTURE

When the sails are not in use, they should preferably be folded or rolled while they are dry and free of salt. This is not always possible while underway, but make sure that the sails are rinsed clean of salt and dried when stored for a longer period of time. A good method for removing salt and dirt from sails is to hoist them on a quiet day while the boat is moored, hose them down with fresh water and let them dry. Rolling the sails are easier to maintain than folding. If you fold, do not fold the sails in the same place every time. Permanent, sharp folds damage sails more than temporary, light folds.

+ REMOVE MOULD AND MILDEW EARLY

If you find mould or mildew on a sail, it is important to act quickly. It can be removed in the initial phase, while it is only superficial – but once it has spread and become saturated, it is very difficult to get rid of. Chlorine is the most effective weapon, but use a weak solution. Chlorine must never be used on nylon (spinnaker cloth) or sails made of aramid fibers (Kevlar, Twaron, Technora). These sails will be destroyed! Hard scrubbing does not help and will only damage the sail. Difficult areas can be submerged in a light chlorine solution for a day. Be sure to rinse well with water. If this does not work, abandon the project. Further attempts will most likely do more harm than good.

+ FIND SMALL PROBLEMS EARLY

Large tears and serious damage to the sails can be prevented if dealt with before they are allowed to develop. Inspect seams at critical locations regularly. In headsails this will particularly be in areas that touches spreaders and stanchions. In the mainsail the critical spots are batten pockets and areas where the sail can rest against spreaders. Make sure the battens are properly attached and secured in their pockets. You can tape or sew small tears or holes yourself in the course of the season.

+ HAVING THE SAILS CHECKED BY A SAILMAKER OVER THE WINTER IS USUALLY A GOOD INVESTMENT.

07

SAIL TRIM - THE BASICS

The basics 99
What is sail trim? 101
5 key elements 102
Angle of attack 104
Heel angle 106
Twist 108
Vortex 110
Balance 112
Fullness 114

WHAT IS SAIL TRIM?

Sail trim is the art of optimising sail profile. The objective partly is to preserve the optimum shape created by the sailmaker and partly to adapt the sail to the actual conditions on the water.

Of course, sail trim influences boat speed. In almost every boat sail trim is the biggest single factor when it comes to influencing speed. But the way we trim the sails also has an impact on other things. If the sails are trimmed correctly, the boat will turn into a more comfortable and pleasant place to be. It will heel less, balance and weather helm will improve and the boat will move better in the waves. Wear and strain on sails, rigging , steering gear and crew will also be reduced.

VARIATIONS OF THE SAME BASIC PROFILE
To the untrained eye the profile of a sail seems to be the same all the time. However, different wind angles and wind speeds, different types of sailing and different types of boats require slightly different profiles. These changes may be fairly small, but they can have a significant impact on how the boat sails. In other words, it really makes a difference when you adapt the sail profile, especially with shifting wind angles and wind speeds. Another major aspect of sail trim is to compensate for the fact that the sail profile is being manipulated and deformed by the varying forces acting on the rig and sails.

CONFIGURING THE DEEPEST POINT
The adjustments you can make are rather small and there are actually not so many different things to change in a sail.

Often it helps to make the sails a little deeper, or slightly flatter. The deepest point of the sail can be moved slightly forwards or backwards, or the twist can be adjusted in order to regulate how much the leech opens up. These are your basic options.

TRIM FOR THE CONDITIONS
It sounds simple and in essence it is. What is not so simple, however, is to get the exact changes you want and avoid making changes that you do not want. A rig has many strings to play on and one change often affects the sails in more ways than one. Various rig types, deck fittings and trim lines create opportunities as well as limitations and help to decide what it is practically possible to achieve.

AN ART FORM
The biggest challenge is to know what trim changes will give the best outcome in any given situation.

This is an art form no sailor will ever fully master. The combination of possibilities is almost endless and constantly changing. Sail trim is something you can play and experiment with your whole life. It is a fascinating process and you will find that your understanding of the dynamics at play will continually evolve and improve. The feeling of getting it right and having the boat move along effortlessly and in full balance is one of the true highlights of sailing.

Remember that it is neither a criminal offense or dangerous in any way to sail with sails that are not optimally trimmed. There is always room for improvement somewhere. So keep experimenting and keep learning. This applies no matter what your level of experience might be.

5 KEY ELEMENTS

What are the most important elements – those that really matter? Here is a list of five key elements, with a few keywords. On the following pages each point is further explained.

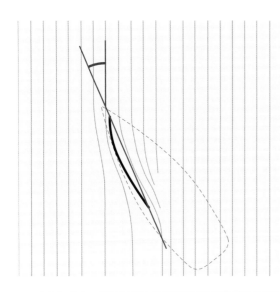

1 Angle of attack

Is the sail deflecting as much air as possible, without stalling? This is what creates the driving force (lift). The angle of attack is mainly regulated by the sheets, along with how the helmsman is steering. This is why the sheets are your most important tools when trimming the sails and why a good helmsman is worth his or her weight in gold. More on page 104.

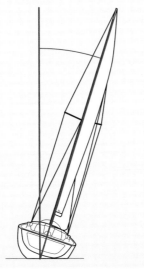

2 Heel angle

Are the forces in the rig adapted to the stability of the boat? The boat is designed with a certain amount of stability (righting moment) and can handle a certain amount of lateral force from the sails. When the wind is forward of the beam, the angle of heel will indicate the amount of force in the sails. This can of course be controlled by changing the sail area, but also by trimming the sails. There is a lot to be gained from using the boat's righting moment as effectively as possible and continually working on the heel angle as wind conditions change. More on page 106.

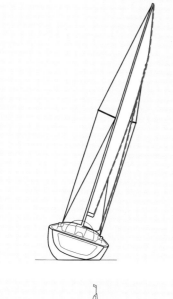

3   Twist

Does the leech in the upper part of the sail open or close optimally for the prevailing conditions? If a sail is to develop the right amount of power all the way up to the head, it has to twist – the gradual falling out of the leech has to be controlled. Twist to some degree is necessary if apparent wind angle is to be constant all the way up. Twist also affects heel and balance. How much the leech is open or closed is primarily controlled by moving the sheeting angle and adjusting the sheets. This really is the key element in sail trim! More on page 108.

4   Balance

Are the forces working in the headsail and mainsail balanced in a way that makes the boat sail straight, without much helm correction? If the boat is not well balanced, the rudder will act as a brake and there will be unnecessary strains and loads in the rig and steering system. Balance can be adjusted in many ways: Sheeting, fullness and twist in the sails, mast trim, heel and several other things. More on page 112.

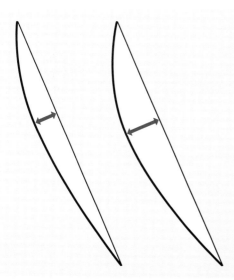

5   Fullness

Is the amount of fullness and the position of the deepest part of the sail right for the conditions? At sharp wind angles, drag produced in the rig will pull the boat sideways or backwards. The goal is to maximise the component that drives the boat forwards (thrust) and minimise side force and drag. Both fullness and position of the deepest part of the sail has to be trimmed according to wind speed, wind angle, waves etc. Trimming these parameters is done with halyard, outhaul, mast bend, forestay tension and several other things. More on page 114.

ANGLE OF ATTACK

Are you giving it full throttle?

This is the engine room, where the horsepower is created. There is an optimal angle of attack and it can be found by using the sheets actively – combined with precise steering. You can feel it when it is right.

Angle of attack can be several things, but in this book we will most often refer to angle of attack as the angle between the sail and the wind. Since we are talking about the engine itself, this is of course an important concept. But let's start by accurately defining the sail's angle of attack (opposite side).

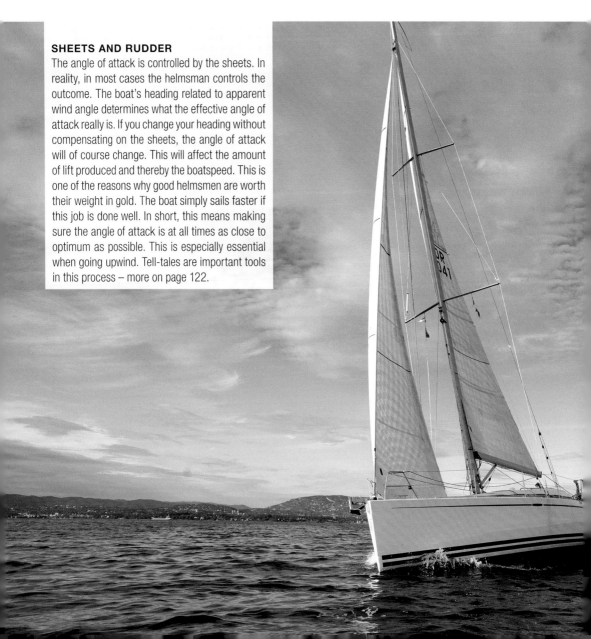

SHEETS AND RUDDER

The angle of attack is controlled by the sheets. In reality, in most cases the helmsman controls the outcome. The boat's heading related to apparent wind angle determines what the effective angle of attack really is. If you change your heading without compensating on the sheets, the angle of attack will of course change. This will affect the amount of lift produced and thereby the boatspeed. This is one of the reasons why good helmsmen are worth their weight in gold. The boat simply sails faster if this job is done well. In short, this means making sure the angle of attack is at all times as close to optimum as possible. This is especially essential when going upwind. Tell-tales are important tools in this process – more on page 122.

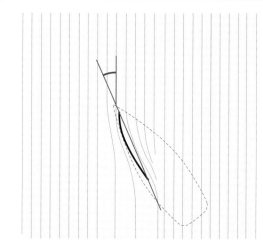

MAXIMUM DEFLECTION WITHOUT STALLING

The angle of attack is defined as the angle between the sail (more precisely the chord, a straight line between the luff and the leech) and the wind (apparent wind angle). The sail's angle of attack determines how much lift the sail produces. There is always a certain angle of attack where maximum lift is produced. At this point, the sail deflects so much air, that just a little more camber or angle of attack would initiate stalling.

The angle of attack should in most cases be the same all the way up the sail and this requires twist. More on page 108.

It goes without saying that the gains can be very big, if you can find a trim that enables the sail to be as close to this optimum angle of attack as possible at all times. This is simply what makes the boat move.

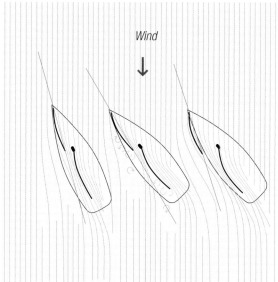

Wind

The boat to the left sails perfectly upwind – pointing close to the wind, but still with laminar air flow and sufficient boat speed. This boat is reasonably fast and points well.
The boat in the middle steers lower to the wind, but with the same trim as the boat to the left. The sails are not adjusted to adapt the angle of attack to the different heading. The sails are stalling. This boat is not fast and does not point well.
The boat to the right has the same heading as the boat in the middle, but this boat has eased the sheets and adapted the angle of attack to the heading. This boat is fast – but does not point well.

HEEL ANGLE

Keelboats are designed for an optimum heel angle. Sails and trim must be adapted to balance side forces and righting moment.

Keelboats have to heel before the keel starts to deliver stability. A keel hanging straight down is dead weight. It just pulls the boat down. The hull itself provides some form stability, but quite quickly more is required. On a keelboat this essential stability comes from the keel, which is why it's called a keelboat. But again: The keel will only work as a stabilising factor when the boat heels (more about the significance of the keel on page 55).

STABILITY AND SPEED

The closer to the wind a boat is heading, the more lateral forces are created by the sails. This demands increasingly more stability from the boat. The righting moment delivered by the keel allows you to generate more power from your sails, which translates to speed. In other words, stability means potential for speed. In the real world, displacement boats will soon hit their theoretical hull speed, which limits how much speed the boat can actually achieve. When the boat reaches hull speed, added power will not help much. Planing boats (dinghies, sports boats, multihulls) can get considerably more value from greater stability and larger sails than displacement boats. The main advantage of high stability and large sails on a displacement boat, is that it reaches hull speed easier, that is in less wind. In other words: Displacement boats with high stability and large sail areas will not necessarily be faster. They will reach maximum speed more often.

TOO MUCH HEEL – NEGATIVE FACTORS

The more the boat heels, the more righting moment is generated from the keel, theoretically up to a heel angle of 90 degrees. Needless to say, it would be both inefficient and uncomfortable to be heeling at 90 degrees. The mast would be almost in the water and the sails would become inefficient. The rudder would stop working too.

However, the negative factors of too much heeling is noticeable long before the boat is heeled to 90 degrees. The performance of most boats will suffer considerably already at 30 degrees heel. Effective mast height is reduced and the wind flows in increasingly distorted angles across the sails. The rudder (especially if there is only one) is lifted closer to the surface. The shape of the submerged part of the hull also gets increasingly distorted and this has a negative effect on balance and boat speed (much depends on boat design, though). Drift will also increase.

Is there a balanced relation between stability and side forces?

NOT ENOUGH HEEL – NEGATIVE FACTORS

Not enough heel, however, reveals that there is not enough power in the sails to achieve maximum thrust. You are not using the righting moment to create speed. This does not apply downwind of course, where the sails will not create side forces to the same degree. On any other point of sail, heeling below optimum heel angle means that you need more pressure in the sails to reach target speed. Most commonly it simply means that there is not enough wind, which is hard to change, but then at least make sure to have as much sail up as possible and to trim the sails and rig for maximum power.

18-20 DEGREES?

Every boat has an optimal angle of heel, where the positive effects of righting moment and form stability are greater than the negative effects of heeling.

This optimum angle varies from boat to boat. Narrow classic boats with long keels and overhangs can take quite a bit of heeling, while beamy flat-bottomed, lightweight boats normally should be sailed more upright. This rule of thumb does not hold completely true for modern performance boats. They are often designed to be faster upwind with a fairly large angle of heel, maybe even 30 degrees or more. In a normal cruiser/racer, however, the optimal angle of heel is often somewhere between 18 and 20 degrees. At this heel angle, the keel will deliver a reasonably good righting moment and the negative consequences of heeling are well within the design parameters.

WHAT IS YOUR HEEL ANGLE?

With this in mind, it is obviously a good idea to know the optimal heeling angle for your particular boat. Talk to people who have experience with the boat in question, or contact the yard or the design office. You can also experiment on your own, by keeping an eye on the log while trying out different sail configurations and trim setups with different heel angles. Pay particular attention to weather helm as heel angle changes (more on balance and helm pressure on page 112).

TRIM TO HEEL ANGLE

When performance is lacking, it might be a good idea to start with an assessment of the balance between stability and side forces. Check the heel angle! Quite often there is much to be gained here.

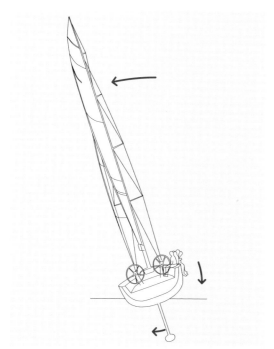

Keel and crew weight (moveable ballast) counteracts heel.

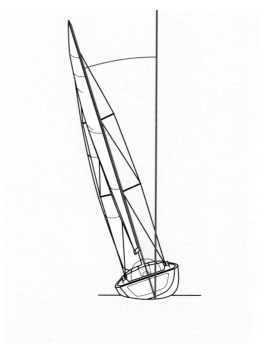

Heel angle in a lot of sailing boats should be 18-20 degrees, to activate righting moment from the keel in the most efficient way.

TWIST

What is twist actually – and why do we need it? That is a very interesting story. There are more things at play than you might think.

If you look at a sail from behind, you will see that the leech opens up more in the top than lower in the sail. That is what we call twist. The sail is twisted in the vertical direction.

Twist happens all by itself, because of the way sails and rigs are designed. All it takes is a little wind and the sail will begin to twist. It is a result of leech tension and this is what we are working with when we are trimming to control twist.

SEVERAL REASONS

Initially it may seem illogical that twist is a positive factor at all. Apparently it means giving the sail a different angle of attack in the top than in the bottom. But as we shall see over the next pages, conditions are not the same all over the sail. In some situations, twist is a way of getting rid of excess power in the sail. In other situations it is an attempt to create a laminar air flow over as much of the sail area as possible.

WHY LIMIT TWIST?

Twist is not necessarily a good thing. More is not always better. Leech tension helps the sail deflect air and create lift. If the sail twists too much, it will not produce enough power. A sail with exaggerated twist will not be able to point very close to the wind – upwind, the whole sail plan has to be trimmed close to the centerline. Finding the exact right amount of twist is a fine line. Not too much, but not too little.

FOUR REASONS TO TWIST THE SAIL

1. Twist is an effective way to get rid of excess power in the sails. When wind speed reaches a point where the boat's stability no longer can handle full power from the sails, you need to remove some of it. Since the top of the sail heels the boat the most, it makes sense to start reducing side force at the top. In lighter conditions, when the boat's stability can handle it, you can "connect" the head of the sail by reducing twist and generate more power.

2. Wind speed at sea level is lower than higher up in the air. The reason is friction between the water surface and the wind. In other words – the wind is always stronger in the top of the sail. But the whole rig moves in the same speed. As a consequence, induced wind becomes a relatively larger component at the bottom of the sail. This means that apparent wind angle will be more on the nose as you get lower. Roughly speaking (and with a little exaggeration),when the low part of the sail is sailing close hauled, the top is close reaching. If the sail's angle of attack is to be constant all the way from the bottom to the top, it has to twist to some degree to adapt to this vertical change of apparent wind angle.

3. Twist does not only affect power, but also balance. You can increase the force in a sail by reducing twist and you can reduce the force in a sail by opening the leech. This can be a way of moving the center of effort. A twisted mainsail will for example move the center of effort forwards and reduce weather helm. The thrust from a twisted top is also directed more forwards.

4. The wind doesn't only move in a straight line from luff to leech. The airflow is forced upwards around the top of the sail and downwards below the bottom of the sail. This causes turbulence and resistance (drag) – especially at the top. Twist helps. See more about vortex on the following pages.

HOW TO CONTROL TWIST?

Upwind, twist is mostly controlled by sheeting point and sheeting. Headsail twist is trimmed with the jib car and the jib sheets. Mainsail twist is trimmed with the traveller and the mainsheet. In addition, all sails will open more in the leech when luff tension is increased. In other words: The leech will open when the halyard is tightened. Reaching and running, twist in the mainsail is controlled by the kick. Telltales are the best tool for assessing twist. More on telltales on page 122.

Good mainsail twist for light/medium wind. In strong winds the sail should twist even more.

Not enough twist in the main. The top of the sail stalls and the boat loses speed.

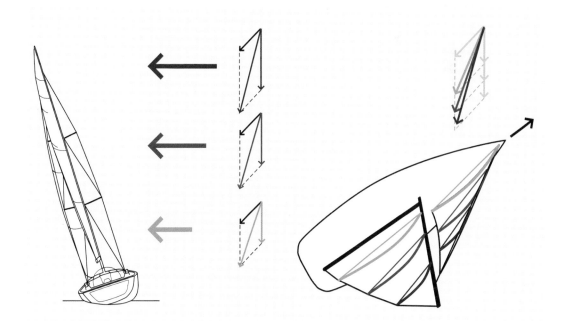

On the right you can see apparent wind from all three diagrams put together (the effect is exaggerated on the illustration). Twist is adapted to the change in wind direction. On the right you can see how the wind is weaker at the bottom of the sail (shorter arrow) than at the top (long arrow). This affects apparent wind angle.
To sum it up: Low in the sail, the wind is weaker and the direction is more on the nose (yellow arrow). High in the sail, the wind is stronger and more from the side (brown arrow).

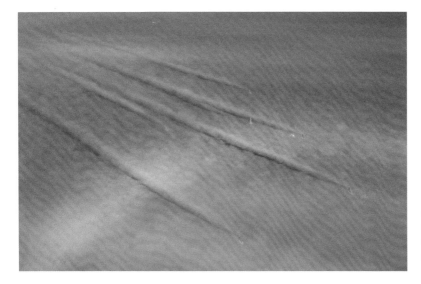

It is very rare to see vortex appear visibly this way. Participants in Volvo Ocean Race are setting out from Cape Town

THE INVISIBLE TORNADO
VORTEX

Airflow on the windward side is bent vertically around the top and bottom of the sail. This creates a "tornado" known as a vortex. This creates a lot of resistance and can be reduced by twisting the sail. But it comes with a price.

It would be natural to assume that the entire air flow over the sail moves horizontally, from the luff to the leech. But it is not that simple.

As the pressure is different on either side of the sail, the wind will try to move from the high pressure side (windward) towards the low pressure side (leeward). This means that the airflow across the sail follows a slightly different pattern than you might initially think. Some of these patterns are very unstable and can change very quickly. Speed, wind angle, heel angle and not the least sail trim are some of the factors that affect airflow patterns. To some degree, turbulence and disruptions of the laminar air flow will develop because of this vertical component, especially around the top of the sails. This often affects the efficiency of the sails quite a lot.

VERTICAL AIRFLOW
On the windward side, the airflow in the top and bottom sections of the sail is bent vertically, in an attempt to equalise the pressure difference. The airflow on the leeward side will to some extent be bent towards the center of the sail, where the pressure is lower.

Under the foot of the sail the high pressure airflow from the windward side will equalise with the low pressure airflow from the leeward side. This creates an eddie, a kind of tornado which is pulled backwards with the wind and creates drag or resistance. This little tornado is called a vortex. If the sail is trimmed with a flat foot, the pressure difference between the windward and leeward side of the sail is reduced. This promotes horizontal airflow and less air tends to flow downwards. The consequence is less vortex production and that is one of the reasons why it is faster to sail upwind with a pretty flat foot.

VORTEX
At the head of the sail things gets more complicated. The triangular shape itself creates problems in terms of more turbulent airflow and the fact that the boat is heeling doesn't help the situation either. The more the boat heels, the more vertically oriented the airflow becomes all over the sail. Heeling means less vortex problems at the bottom of the sail, but more at the top. In addition, there is the influence from waves: Every movement of the boat is exaggerated in the top of the rig. As you see, there are several reasons why drag from vortex and turbulent air in general can be big problems at the head of the sail. Fortunately, this is also the easiest place to do something about it. See the next page.

Quite a bit of twist in the sail. Less lift produced at the head of the mainsail, but also less vortices.

Not much twist in the sail. More lift produced at the head of the mainsail, but drag from vortices increases too.

What is vortex?

A vortex is formed at the end of every lift-producing foil, whether we are talking about airplane wings, rudders, keels or sails. In short it is an eddie, a swirl or tornado, developing when the flow on the high pressure side (windward) is deflected in an attempt to equalise pressure. Air (or water) is turning towards the end of the profile, to unite with the flow on the low pressure side (leeward). A collision happens when the two airflows meet, each with a different speed. Vortex causes drag, a force pulling the boat sideways and backwards.

As long as a sail creates lift and thereby a pressure difference, it will to some degree also develop vortices, especially at the head of the sail. If the head of the sail is completely flat, there is no pressure difference – and no vortex. The same happens if the angle of attack is zero: No pressure difference, no vortex. But in both cases the consequence is that no lift is produced either.

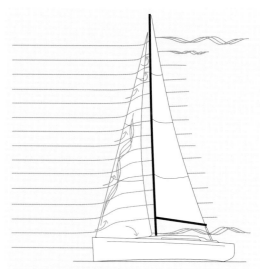

The forestay meets the wind at an angle – the lowest part comes first. The wind constantly seeks to equalise pressure and is therefore deflected upward, towards the head. When close hauled, a swirl is easily created, moving up along the luff on the windward side. This can be revealed by telltales.

WHAT CAN WE DO?

As vortex is a consequence of lift, we can reduce vortices by trimming the top and bottom of the sail flatter. A flatter profile provides less lift – and less vortex. We can also reduce the angle of attack. A more neutral angle of attack will also produce less lift and less vortex.

Twist is a way of reducing angle of attack in the top section of the sail and usually a flatter profile will follow. A flat, twisted top produces little or no vortex.

When you increase twist, you produce less lift, but also less drag. The center of effort moves forward, affecting the balance in the boat. The lift produced in the head of the sail, even though reduced, is directed more forward and heeling is significantly reduced. In certain conditions (mostly light/medium wind) it will be more important to increase power than to reduce vortices. In those conditions you gain by sailing with less twist, in spite of more vortex drag.

HOW MUCH WEATHER HELM?
BALANCE

Try to make the forces working in the headsail and mainsail balance until the boat sails straight ahead, almost by itself. The rudder acts as a brake every time it is used – but you still want a small amount of weather helm.

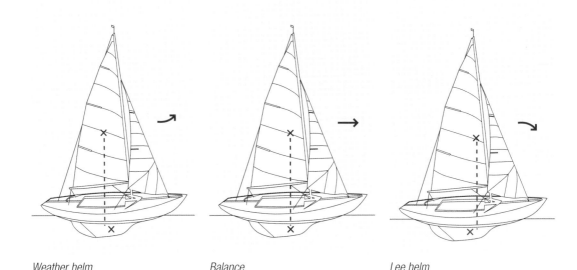

Weather helm   *Balance*   *Lee helm*

CENTER OF EFFORT AND LATERAL RESISTANCE

In theory, the collected forces working on the sails can be located to one point, called the center of effort. In the same way, the forces working on the keel and rudder can be located to another point, called the center of lateral resistance. So, we have two sets of forces collected and located to two single points, one in the sails and one under water. Only if these two points are on the same vertical axis can the boat be in perfect balance.

When the center of effort moves aft of the center of lateral resistance, the boat will be forced against the prevailing wind direction. This is what we call weather helm.

When the center of effort moves ahead of the center of lateral resistance, the boat will be forced away from the prevailing wind direction. This is what we call lee helm.

With a few exceptions the keel and rudder can't be moved. Heel angle affects center of lateral resistance, but apart from controlling heel angle, there is not much you can do to change it. We can however affect the center of effort quite a lot by sail trim. The obvious example is letting out a little on the main sheet or traveller, to reduce weather helm.

RELEASE THE BRAKE

If you find yourself constantly pulling on the tiller or steering wheel to counteract weather or lee helm, it is pretty much the same thing as driving a car with the handbrake on. The rudder blade will be angled against the water flow, creating resistance. If the boat is "hard on the helm," it is a very clear message: One or several trim functions should be changed. How this can be done specifically will be dealt with later in this book, but the short version is simple: You have to reduce or increase the pressure in either mainsail or headsail – or move the center of effort within each sail. Heel angle is also an important factor.

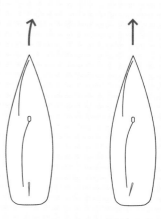

SLIGHT WEATHER HELM OK

Many experienced sailors like to trim the boat to induce slight weather helm. This makes it a little easier to sail upwind, pointing as high as possible against the wind. A tendency to weather helm gives the helmsman a good "feel" of the boat. There is a direct contact with the rudder. Ideally, you should be able to let go of the helm and the boat should slowly round up against the wind. Rudder profiles are designed to work optimally with a slight angle to the water flow, this is another reason for trimming towards a preference for slight weather helm.

WHAT IS A GOOD PROFILE?
FULLNESS

A sail profile looks like a circular arc with the rear end pressed flat. Broadly speaking, the deepest point should be 40-50% from the luff and depth should range between 10 and 20% of the width of the sail. But this is a gross generalisation. In reality, no sail profile works well for all purposes.

Let's look at a hypothetical boat sailing upwind. We assume that the angle of attack is set right. The sail is deflecting as much wind as possible, but still with laminar airflow on both sides of the sail. Heel angle is just right, meaning that the sail area and the power in the sails are well adapted to the boat's stability. Both headsail and mainsail are twisting the way they should, indicating that the angle of attack is constant all the way up the sails. The balance is adjusted until there is a slight tendency to weather helm.

What can we do to optimise things further?

SMALL BUT SIGNIFICANT DIFFERENCES
The last of the five core points in this chapter is about the shape and size of the sail profile. You need to make sure that the deepest point in the sail is exactly where it should be and that the fullness of the sail fits the conditions. To know how to do this, it is necessary to take a closer look at what a good sail profile really is.

The answer is that it depends. The optimal profile in light breeze and strong winds is not the same. A good profile upwind is not the same as a good profile downwind. There are differences between the mainsail and headsail and it is also necessary to adapt the sail profile to different types of boats, as well as wind and wave conditions. For the untrained eye these may seem like small variations, but they are quite noticeable when applied.

WRINKLES ARE NOT IMPORTANT
Exactly what differences are we really talking about? Well, apart from the fullness and position of the deepest point of the sail, it is the shape of the entrance (how rounded or flat the forward part of the sail is) and the tension in the leech (that is, the twist in the sail). Wrinkles and folds in the sail are of little or no significance however. What matters is that the profile is properly adapted to the conditions. In light winds, for instance, it might be quite good to sail with

horizontal stripes in the luff. It just means that the tension in the luff is low, which provides a deeper, more powerful sail with depth located relatively far back. That is exactly what we need in those conditions. Strong winds demand something else, Here the luff and foot must be tightened up much harder, to provide a flatter, more open sail profile with the deepest point moved forwards. It's all about the profile, not how smooth the sail looks!

GREATER DEPTH UP TOP?
Looking at how fullness is distributed vertically (that is upwards in the sails), what you need depends on the wind speed. In light winds increasing fullness as you look higher up in the sails is a good thing. We want maximum power and the wind is stronger further away from the water surface. As the wind increases it becomes necessary to limit heel and this is done most effectively by making sure that the top and mid-section of the sail becomes a little flatter. This need increases in strong winds when the sails need to be flattened and twisted even more.

THE OPPOSITE OF WHAT WE NEED
The pressure in the sails increases proportionally to the square of the wind speed. So when the wind speed increases from 10 to 20 knots, the loads working on the rig is four times higher. All sailcloth stretches to a certain degree and masts will bend. Rope and even steel wire will stretch. Under those circumstances it is not surprising that sail profile will change too. Unfortunately the changes are not for the better. Strong winds will distort the sail shape if you let them. If nothing is done, the sails will become fuller and the deepest point will move aft. The result is more power and at sharp wind angles the forces will to a large degree be directed sideways and backwards. If sail trim is not adapted to the conditions, you will get the opposite of what you need.

COUNTERACT DEFORMATION

The main focus should be to make sure sails do not distort or lose the profile they were designed to have – in other words, to compensate for distortion. Very often, the first item on the agenda is to prevent the depth of the sail from moving aft as the wind increases. Chapter 8 will go more in detail on how to work with the different trim tools to shape the sails the way we need in different situations.

Mainsail: Flatter bottom section than head, deepest point almost in the middle, twist adapted to conditions.

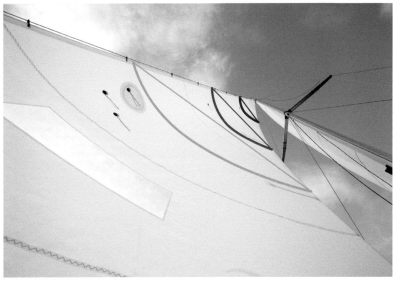

Headsail: A little more fullness than the mainsail, the deepest point pulled more forward. Also some more twist.

Vind

HOW TO CHANGE FULLNESS

Depth or fullness is basically changed by changing the distance between the luff and the leech. Imagine holding a sheet of paper and alternately stretching and bending it. In a rig, a similar thing can be done by providing more mast bend and a tighter sheet (flatter mainsail), or by providing a tighter forestay and a tighter sheet (flatter headsail). To increase fullness you would of course do the opposite.

HOW TO CHANGE THE POSITION OF THE DEEPEST POINT

The position of the deepest point can be changed a couple of different ways. One is by stretching or straightening the luff curve. In the headsail, luff curve is controlled by forestay tension (sag). In the mainsail, luff curve is controlled by mast bend. In both cases, less luff curve will move the deepest point forward and increase fullness. In both cases, backstay is the principal tool.

Another way to manipulate the deepest point in the sail, is changing luff tension. The tool for that is the halyard. Tighter halyard will move the deepest point forward (and flatten the sail).

Where the correct position of the deepest point is, will vary with circumstances. There is no single correct position. Try to trim mast bend and sag until the luff curve provides the sails with the amount of power and pointing ability that is right for the circumstances. Try to achieve a good balance between tension in luff, foot and leech.

For a better understanding of these dynamics, check the section on how sail profile is built into the sails (page 68).

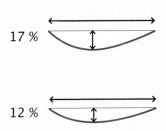

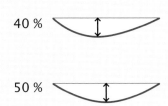

The size of the depth is measured in percent of the length of the chord.
Here two examples.

The position of the deepest point is measured by finding the point on the chord where the sail is deepest and calculating the distance to the luff as a percentage of the total length of the chord.

HOW DEEP SHOULD THE SAIL BE?

New apps with photo analysis can be used to measure both the size of the depth and the position of the deepest point. It may be useful to be aware of what these measurements should be. We have already looked at measurements for depth position (percentage of sail width measured from the luff, normally 40-50%) - but what about figuring out exactly how full the sail really is?

The depth of the sail is also measured using a percentage of the chord. To calculate the number, sailmakers simply measure the distance between the chord and the deepest point of the sail in a right angle (look at the illustration). Then they calculate what percent this distance is, related to the length of the chord.

In light winds (less than 8 knots), the depth should be 13 -16%, preferably a little more for headsails. In moderate winds (8-16 knots) depth should be reduced to 11-13% and in strong winds (over 16 knots) a good objective is 9-12%.

In addition, you would typically want to move the deepest point slightly forward with increasing wind, more so in the headsail than in the mainsail.

Horizontal lines taped in the sail at different levels are useful tools to assess both maximum depth and the position of the deepest point. In a cross cut sail the seams are more or less horizontal and can also be used for a visual check of profile.

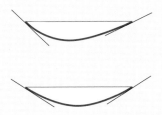

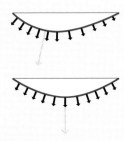

Entry angle and exit angle are decided in design programs before the sail is built. But you can change these parameters through sail trim.

Here you can see how the position of the deepest point affects the direction of lift force. Depth pulled forwards (towards the luff) will give more thrust ahead. Depth further aft (towards the leech) will improve pointing ability.

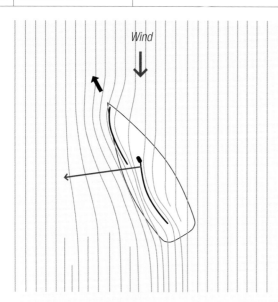

Boat and rig, especially the maninsail, affect the wind in front of the boat. This provides the headsail with a more favorable wind direction.

Notice also how the wind moves around both sails as if it was one foil. Red arrow shows lift direction.

THE WIND WANTS TO AVOID THE BOAT

The headsail comes first and generally has free, undisturbed wind. In addition, it benefits from a certain deflection of the air flow created in front of the boat or forestay – look at the illustration. The effect is similar to what happens in front of a rock in a stream: The flow of water in front of the stone is deflected and bent already before it reaches the stone. The boat and rig – especially the mainsail – moves into the wind with a certain mass, forcing the wind in front of it to change direction.

In a sailboat going upwind, the angle to the wind means that this deflection is always the same. The wind will deflect in the direction of least resistance, in this case to leeward. That is what you would normally call a lift – a wind shift that allows you to point higher. When close hauled, the headsail receives an airflow that is quite a few degrees more favourable than the mainsail gets.

This effect makes it possible to move the deepest point of the headsail further forward (up to 35-40%, sometimes even further) and optimise the thrust of the sail – that is, the force from the sail that moves the boat forward. Because the airflow meets the luff at a wider angle, the entrance can be rounder than the entry of the mainsail.

All in all it can be considered to be a wind shift that is always there, just ahead of the boat. Large and heavy boats with powerful rigs carry more mass and will generate much more of this deflection than smaller boats.

SEE BOTH SAILS AS ONE FOIL

Try to view the headsail and mainsail as one single "foil", with a slot between the two parts. That is how it actually works. Looking at it from the perspective of the wind, the headsail comes first and then the main – but we are talking about a combined, lift-producing package. The two sails do not perform quite the same job and for that reason, they should not have quite the same profile.

The headsail can be seen as a more or less fixed entry, producing a lot of lift, while the mainsail can be seen more like a flap, used to adjust pressure and balance – much the same way as the flaps on airplane wings.

Going upwind, the rear part of the headsail should be very flat and usually twisted more than the mainsail. The reason is primarily that the airflow behind the headsail should avoid being directed into the leeward side of the mainsail. You want to maintain a fairly even gap between the two sails. This becomes increasingly important as the headsail overlap increases. This is why large genoas should twist more than smaller headsails.

The entry of the mainsail should be flatter than the entry of the headsail. The leeward side of the mainsail receives a compressed airflow from the gap between the two sails and is greatly affected by the headsail. The deepest point in the mainsail should usually be further aft: Close to 50% in light wind and a little less in a breeze. Looking at the sails in this perspective, it becomes clear how important it is to move the boom all the way up to the centerline to provide pointing ability. Some boats even trim it to windward (if stability allows).

08

PRACTICAL SAIL TRIM

Using the trim lines 121
Telltales 122
The backstay 124
Forestay sag 126
Mast-bend 127
Running backstays 128
The sheet 130
Headsail halyard 132
Sheeting point 134
Barberhaul/tweaker 136
Mainsail halyard 139
Cunningham 140
Outhaul 141
Kick or boomvang 142
Traveller 144
Mainsheet 146
Mainsail trim in gusts of wind 150
Spinnaker trim features 152
Lines not used for trim 156
Spinnaker topping lift 158
Afterguy 160
Helming and sheetlng 162
Trim features of the gennaker 164
Sheets, tackline, halyard and barberhaul 166

USING THE TRIM LINES

In this chapter we will go through each sail and explain how to trim it. Each sail has a set of trim lines and in this section we will look at how to use each of them.

If you have already been through the previous chapters, you will have the basics required to use the "toolbox". We will be using the tool box in this chapter. It's time to be practical and to the point: What do you actually do and which lines do you pull to make the boat go?

THE MOST COMMON TRIM FUNCTIONS

Boats are of course rigged and equipped differently. We do not always have the same set of tools at our disposal, although a lot of trim lines are common and in use on most boats. But almost every boat has sheets, halyards, some adjustable sheeting points, a kick, an outhaul and very often a few more lines to work with.

THE RIG AND ALL THE SAILS

We will start by looking at what can be done with the rig while under sail. Most often trim control is limited to the backstay, but as you will see, the backstay is the key to trimming both the headsail and the mainsail. Subsequently we will look at the sails one by one and how to use the various trim lines attached to each of them.

TELLTALES

Regrettably the airflow across the sails is invisible. But telltales will tell you.

Telltales are simply small threads attached to the sail, put there to tell tales about the path of the airflow across the surface of the sails. They reveal any tendency to turbulence by flickering. If the airflow is horizontal and laminar (a useful scenario) the telltales will stream straight backwards. On a boat that is helmed properly, with suitably trimmed sails, the telltales should stream straight aft on both sides of the sail.

There are of course some exceptions to this, which we will come back to.

ALONG LUFF AND LEECH

Telltales can be positioned at various levels close to the luff or in the leech. This is where things really happen and this is also where we can influence sail profile the most. Telltales are used slightly differently, depending on whether they are attached forward or aft in the sail. In a stayed rig with headsail and mainsail (the kind of rig that is most relevant in this book) it's appropriate to place three sets of telltales close to the headsail luff: Low, in the middle and high up in the sail. A single telltale in the leech, about three quarters of the way up, is also a good idea.

In the mainsail it will be most beneficial with a telltale placed at the end of the top batten, high in the leech. Placing telltales at the end of all the battens is also useful, the top one being the most important. A set of telltales placed close to the luff, low in the sail, is also advantageous.

REVEAL ANGLE OF ATTACK

Firstly, let us look at what happens in the front of the sail. The telltales along the headsail luff are placed in pairs, one on each side of the sail, around 20-30 cm aft of the luff. From this position the telltales will reveal the airflow's direction into the sail, the so called "entry". This is a particularly interesting place to keep an eye on, because this is where laminar air flow is established. Telltales near the luff indicate the angle of attack – see more about angle of attack on page 104.

If you do have multiple sets of telltales vertically, it is possible to keep track of the entry and angle of attack all the way up towards the top of the sail – see illustration on next page.

REVEAL TWIST

Telltales in the leech have a slightly different function. They reveal the airflow out of the sail, the so called exit. Tur-

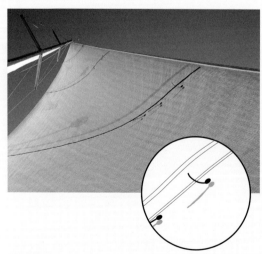

The telltale on the windward side is fluttering. It is signalling that you are steering too high, or that the sheet needs to be tightened. The airflow entering the windward side can't follow the curvature in the sail. The wind is hitting the sail too much from the leeward side. This is the start of luffing, which is very easy to see.

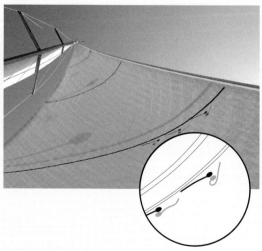

Telltale on the leeward side flutters. It shows that you are steering too low, or that the sheet is too tight. The airflow entering the leeward side of the sail can't follow the curvature in the sail. The wind is hitting the sail too much from the windward side. This is difficult to detect without telltales.

bulence here means that leech tension (which translates to twist, see page 108) is either too high or too low. In essence, telltales in the leech will tell you if the sail closes too much or not. If the leech closes until the sail's stalling point, the telltale will point to leeward, or completely disappear behind the leeward side of the sail. It "hides", so to speak.

The sail twists just right when the top telltale in the leech flutters backwards – but only just. It should preferably have a tendency to hide to leeward some of the time (about 20%). This will tell you that leech tension is sufficient, but not exaggerated. On page 108-111 you will find details about twist and vortex.

EXCEPTIONS

This method applies to most boats, as long as you want maximum power out of the rig. But in strong wind the sails should twist more (downwind excluded), in order to release excess energy. In heavy weather the telltales in the leech should be streaming all the time.

You can even steer or trim briefly until the windward telltales in the front of the headsail starts fluttering. This can be a good way to get through a gust where the boat is temporarily overpowered.

In other circumstances it may be advantageous to close the sails a little extra, just past stalling point – for example in situations where you need brief tactical pointing ability.

The sail has the correct amount of twist. The telltale on the top batten is streaming. But is the sail twisting too much? To know this you need to keep a regular eye on it. This telltale should partially hide in the lee of the sail – about 20% of the time is usually optimal. Also check if the upper batten is parallel to the boom. That is a good sign. Twist in the mainsail is controlled by sheet and traveller. See more on page 108.

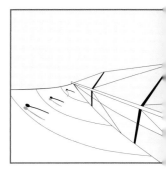

The topmost telltale flutters on the leeward side. This is normally a sign that the sail is not twisted enough. Sheeting point should be moved aft and sheet adjusted accordingly.

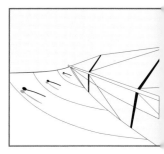

All three sets of telltales along the luff are streaming aft. This means that the sail is twisting correctly and in full power mode. Sheeting point is where it should be and the sheet is trimmed correctly.

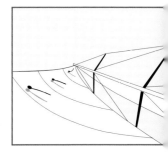

The topmost telltale on windward side flutters. This is normally a sign that the sail is twisting too much. Sheeting point should be moved forward and the sheet adjusted accordingly. In strong wind, this could be a suitable trim.

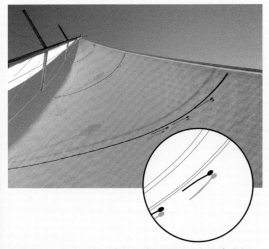

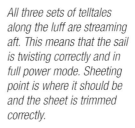

Telltales on both sides of the sail are streaming aft. It is signalling that the sail is pulling optimally: Airflow is entering the sail with a perfect angle of attack and is laminar on both sides.

THE BACKSTAY

Sails are suspended on the rig and of course, this has a major impact on how the sails behave. Sail trim starts with making sure that the rig promotes the efficiency of the sails. The main key while under sail is the backstay.

For most cruisers it will be perfectly adequate to set the rig to a reasonable base trim (as described in Chapter 3) and leave it at that. Sail trim then becomes a matter of using the trim functions that each sail is normally equipped with; halyard, sheet tracks, traveller and so on. The backstay is the only exception.

SPECIAL KNOWLEDGE IN CLASS BOATS

Not all sailors are satisfied with leaving it at that. When racing, it is interesting to try to adapt several parts of the rig trim to the existing wind conditions. This applies in particular to class boats. Generally, in these class environments, a lot of knowledge has been gained when it comes to settings that make that particular type of boat go fast in different conditions. The number of classes and their associated rig trim solutions is vast and there are specific trim guides for all major classes readily available, both online and elsewhere. This book is therefore not the place to look for that type of specialised knowledge.

LONGITUDINAL RIG TENSION

The focus here is on more general things – dynamics that will work for most common keelboats. So what can we normally do with the rig while sailing, as part of our effort to trim the sails?

Actually, not that much! When the mast is set up properly, with good shroud tension, the rest is largely a matter of how hard the rig is tensioned longitudinally. This determines how tight the forestay is and also usually the amount of mast bend. So rig trim under way is only basically about two things. But they are two very important issues!

DIFFERENT WAYS TO TIGHTEN THE FORESTAY

When it comes to longitudinal tension of the rig, there are often several factors at play. On boats with raked or swept spreaders, with the chainplates mounted behind the mast, the shrouds provide a certain tension in the forestay. This is rarely sufficient upwind when the wind pipes up. More tension is needed. In many smaller boats the mainsheet (i.e. the aft part of the mainsail) is an important factor for tension in the forestay. Only a few boats today are set up with running backstays, but if rigged, they are efficient tools, created with this specific purpose in mind.

On most boats, however, there is one thing that really matters when it comes to tensioning the forestay and bending the mast: the backstay.

ACTIVE USE OF BACKSTAY

On racing boats the backstay-tensioner is often placed near the helmsman, to make it accessible from the steering position, or near to the mainsail trimmer who normally sits beside the helmsman. It is used very actively and with small increments, together with the mainsheet and traveller. When the wind picks up, the backstay is used to take excess forces out of the sail. In lighter winds the backstay adjusts the balance between pointing ability and speed, by fine-tuning the tension in the forestay and mast bend. This works best on boats with fractional rigs.

More about luff curve on page 68.

USE OF BABY STAY

Boats with a masthead rig can be rigged with a baby stay to pull the midsection of the mast forward. This makes it easier to bend the mast on this type of rig, where the backstay alone can have a hard time imposing mast bend. The baby stay can usually be adjusted under way. When the mast has a certain pre-bend, the backstay can be tightened further and this will make the process of bending the mast more efficient.

DIFFERENCES UPWIND, REACHING AND DOWNWIND

When we talk about the use of backstay on these pages, it is in relation to upwind or close reaching. Only on sharp wind angles is there a lot to be gained from active and knowledgeable use of the backstay. As you can see in the drawing below, forestay tension is of great importance to the profile in the headsail, while mast curvature is important for the profile of the mainsail. The backstay should be released when sailing downwind or on a broad reach, until the mast reverts to minimum pre-bend and the forestay is relatively slack, giving deeper sails and more power. There is no reason to use the backstay actively downwind. Once you bear off, other things come into play.

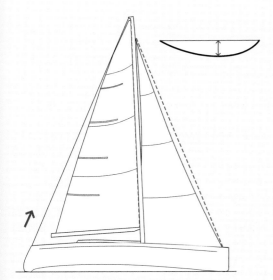

Loose backstay. The mast will straighten, the forestay will be slack. Both sails will become fuller and their deepest points will move forward.

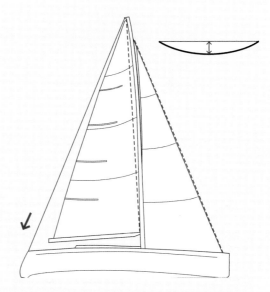

Tight backstay. The mast will bend, the forestay will be tight. Both sails will become flatter and their deepest points will move aft.

FORESTAY SAG

Use the backstay to adjust forestay tension. This will greatly affect the profile of the headsail, making it possible to adapt the sail to different wind and wave conditions.

A tight forestay reduces luff curve and sag in the headsail. The leading edge (entry) becomes flatter. This requires more precise steering and trimming in order to keep the sail from stalling. The whole sail will be flatter and the leech will open slightly. In other words, you will get a more sensitive headsail which produces less power. In return your pointing ability will improve and the sail will produce less side force. This is especially preferable when the wind is up and the sails are producing more power than the stability of the boat can manage. You can reduce heeling and point better – so harden up the backstay in strong winds!

SMALL MARGINS
In a moderate breeze on flat water, a relatively tight forestay can be an advantage, as it gives you good pointing ability. However, it requires a focused helmsman and the margins are tight. Any slight change in the backstay will be noticeable, so this is a balancing act! If the boat feels "dead" and you have a hard time keeping up pace, try easing a touch on the backstay. A little more sag in the forestay might be all you need.

SAG CAN BE GOOD
In light air, especially if it is choppy, you need a little slack in the forestay. More sag gives a rounder entry and the sail does not stall so easily. Sag can be particularly helpful when the bow is being knocked around in the waves. At the same time, the sail will become fuller and produce more power. In return pointing ability will suffer – but in these conditions it is more important to keep up the momentum of the boat.

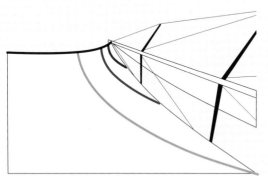

Loose backstay, more sag. Fuller sails, deepest points further forward. Round entry, wider track, more powerful sails. Pointing not so good.

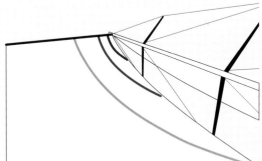

Tight backstay, less sag. Flatter sails, deepest points further aft. Flat entry, narrow track, less powerful sails, but also less lateral force. Points well.

MAST-BEND

Dependant on the type of rig, the backstay can be used to bend the mast. Mast bend has consequences for the profile in the mainsail and can be used actively to adapt the mainsail to changing conditions.

A tight backstay gives you a curved mast – especially on fractional rigs, but also on masthead rigs with a baby stay or double lowers. A curved mast compliments the mast curve built into the sail. The result is a flatter entry as the deepest point is moved aft and the leech opens up. The whole sail flattens, apart from the lower part of the sail, where fullness to a greater extent is governed by the outhaul. All in all you will get a mainsail with the same characteristics as a headsail with a tight forestay, as described on the previous page. If you tighten the backstay really hard (along with the halyard and the outhaul), the mainsail will be well trimmed for rough weather and produce significantly less lateral force.

DEEPER, MORE CLOSED SAIL
If you let out on the backstay, the mast will straighten up. The deepest point of the sail will move forward, the leech closes, the sail will be fuller and the entry rounder. Again,

these characteristics correspond to what happens with the headsail, when the forestay is slackened and sag increases. The sail will be more powerful, but pointing ability suffers and side forces will increase. In light and medium winds the backstay should be slackened gradually as wind speed drops. How much to release will vary as there are major differences between different types of boats, rigs and mainsails. The best way to find out what works is to experiment.

CURVED MAST IN LIGHT WINDS?
As a general rule, the stronger wind, the more the mast should bend. However, in very light winds it may be a good idea to tighten the backstay a little to provide the mast with a certain curvature. A somewhat flatter entry, depth further aft and plenty of twist could actually be a good recipe for an effective upwind trim in light air. Just make sure you can keep the pace!

Loose backstay, straighter mast. More fullness, deepest point further forward. Round entry, wider track and more powerful sail. Pointing not so good.

Tight backstay, more mast bend. Flatter sail, deepest point further aft. Flat entry, narrow track, less powerful sail, less lateral force. Points well.

RUNNING BACKSTAYS

Tighten hard for good trim upwind in a blow.

On a fractional rig with the forestay sitting low on the mast and a slender mast profile, running backstays are required. This is especially the case with in-line spreaders. Running backstays ("runners") have a firm holding effect directly on the forestay and are tightened very hard when sailing upwind in strong winds.

The main difference between the backstay and running backstays is that running backstays are attached to the mast at the same height as the forestay, instead of going all the way up to the masthead.

This difference might not sound so crucial, but on boats where the forestay sits low on the mast (as in 7/8 rigs) it actually means that the running backstays have a different function than the backstay. The running backstay simply takes direct hold of the forestay and tightens it more efficiently. If the running backstay is tightened very hard, it can also contribute to a curved mast – but the backstay does this even better. More about running backstays on page 124.

EASIER WITHOUT
Modern fractional rigs are often constructed with the forestay attached higher i.e. closer to the masthead, with swept spreaders. The mast profile is often stronger, meaning that running backstays are less crucial. Sailing a boat without running backstays is significantly easier and requires fewer hands on deck and a less experienced crew. There are good reasons to avoid running backstays, if you do not absolutely need them and they are consequently rarely found on newer rigs.

TRIMMABLE RIG
Running backstays should be handled with precision. There is less room for error, they require more hands and therefore a larger crew to operate. Many older racers have rigs that are so flimsy that the mast will in fact break if the running backstays are not set correctly. They should be released and tightened to leeward and windward, respectively, every time you tack or jibe, so that only the windward one is active.

In return you get a rig that is very easy to trim. You can get mast bend without losing pressure in the forestay and it is possible to keep the mast curvature at the top of the mast, where it is most needed. Check stays will stabilise and straighten the lower part of the mast. Upwind in a lot of wind, the running backstays are often tightened extremely hard. The greatest effect is seen on the headsail, where you can easily achieve the trim described on page 126.

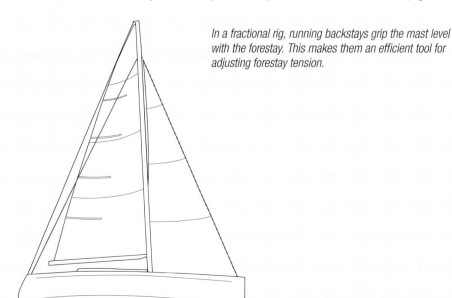

In a fractional rig, running backstays grip the mast level with the forestay. This makes them an efficient tool for adjusting forestay tension.

HEADSAIL TRIM FEATURES

As we have just seen, forestay tension is essential for headsail trim. In addition we have four other options; sheet, halyard, sheeting point and (optionally) barberhaul.

THE SHEET

The sheet adjusts the sail's angle of attack and are your most important trimming tool. Upwind, it is also the main key to the fullness of the headsail.

The profile that the headsail will ultimately achieve when the sheet is tightened or slackened, depends on how the other trim functions are set. Upwind, telltales along the luff are good aids. They indicate if the sail's angle of attack is correct all the way up and down the sail. Upwind they will show whether the helmsman is steering precisely or not. If you want to keep a particular heading (e.g. a close reach), the headsail must be trimmed to the exact course. In a situation like this, the telltales will tell you when the sheet is tightened just right.

A FIST FROM THE SPREADERS

The distance between the sail and leeward spreader (the lower one if there are multiple spreaders) is also a good reference. Upwind, an overlapping genoa is usually OK if this distance corresponds to the width of a fist. Of course this is only a rule of thumb and it is important to remember that the distance is not solely a result of how tight the sheet is, but also a result of how the jib car is trimmed. Only when sheeting point has been correctly set can this reference point be used for trimming the sheet.

The headsail is usually sheeted in until the foot ends up inside the lifelines. It may be necessary to give it a helping hand when the sail is pulled over. Many a headsail has been ripped by stanchions!

TELLTALE IN THE LEECH

A telltale in the leech, about three quarters of the way up can be a valuable tool, when sailing upwind. This telltale should not "hide" in the lee of the sail, but can be allowed to creep behind the leech now and then. Again, we are talking about an interaction with the sheeting point. Twist is a result of a combination between sheet trim and sheeting point.

PERFECT ANGLE OF ATTACK!

Ideally, the sheet should be trimmed for a constant laminar flow and maximum deflection throughout the sail. This is seldom achieved fully, but that is the aim. The sheet regulates the angle of attack and is your main weapon. The more you succeed, the faster you will go, so experiment with small changes. When racing, most ambitious crews will have a trimmer permanently occupied with this job alone.

Many boats today are rigged with relatively high, narrow headsails ("high aspect") and they are particularly sensitive to sheet trim. High aspect jibs should not be trimmed completely flat in the foot and should be trimmed in very small increments.

Common mistakes are sheeting in too hard in light winds and not hard enough in high winds. If in doubt, release the sheet until the sail slightly luffs and tighten again until it sets. No more, no less!

REACHING: A COMPROMISE

Sheeting a headsail on a reach is a matter of compromise. When the sheet is released and trimmed to a deeper wind angle, the head of the sail will twist a lot – usually too much. It helps to move the sheeting point forward, but on a broad reach this does not solve the problem. The genoa track sits too close to the boat's centerline for the jib sheet angle to be wide enough when sailing off the wind. The top of the sail will inevitably be too loosely sheeted when the bottom section of the sail is sheeted correctly. Likewise, when the top of the sail is sheeted correctly, the bottom section will inevitably become too tight. As the lower part of the sail is so much larger than the top, it is probably best to adjust the sheet accordingly. The barberhaul will help, but not completely, at least not on a monohull – more about this dilemma on page 136.

DEAD DOWNWIND: USE A POLED OUT HEADSAIL

Effective sheeting of the headsail dead downwind is difficult bordering on hopeless, unless you "goose wing", i.e.sheet to windward – preferably using a spinnaker pole or whisker pole. If the headsail is set on the same side as the mainsail, it will end up in the wind shadow of the mainsail and have little or no effect. A poled out headsail should be let out far enough to put the sail more or less perpendicular to the boat. If the wind is up, be a little careful not to ease the sail too far forward – it will twist out and lift in gusts and if the head of the sail ends up ahead of the bow, it may cause a broach. In light winds however, it is a good idea to ease the headsail far ahead. The forestay should be as loose as possible.

HEADSAIL

HALYARD

Simply put: the stronger the wind, the tighter the halyard. The lighter the breeze, the looser the halyard. But what is actually happening with the sail?

The halyard regulates luff tension, which is absolutely essential for the profile of the sail. Tightening the halyard moves the deepest point of the sail forward, creating a more rounded front, opening the leech. Basically, you are simply moving sailcloth forward. The profile becomes flatter throughout the aft part of the sail. The rule of thumb is quite simple: The more it blows, the tighter the halyard.

RETURN TO DESIGN PROFILE

As the wind increases, the sail is deformed by the forces acting upon it. This is particularly true when sailing upwind, so close hauled it is very important to counteract this "destruction" of the sail profile. One of the most effective ways to do this is to tighten the halyard. You simply recreate the profile the sail was originally designed to have.

ROUNDED ENTRANCE IN HIGH SEAS

If the wind gets up even more and you have to get rid of excess power in the sail, the halyard should be tightened even further. The boat will sail better in rough weather with a rounded entry in the headsail, the deepest point further forward and a flat aft area of the sail, thus providing a leech that is a little more open. In big waves, the bow will be moving about in all directions. In these conditions, a rounded entry in the jib will be particularly useful in stopping the sail from stalling as the angle of attack constantly changes. A flat entry, resulting from a looser halyard, means the sail has to engage the wind at a narrow and very precise angle. This requires calm water and a concentrated helmsman.

SLACK HALYARD IN LIGHT WINDS

If you need a more powerful headsail, try to ease the halyard a little, possibly along with the backstay, providing more sag. In extremely light winds, consider maximising power in the headsail by sailing with a very loose halyard. Horizontal streaks or folds from the luff don't matter. You will get a deeper sail with a flat entry, the deepest point located relatively far aft and a leech that doesn't open up too much, losing valuable power.

You always need significantly more luff tension going upwind compared to downwind. As a rule, the halyard should be eased quite a lot when on a broad reach and sailing downwind.

INTERACTION BETWEEN SAG AND HALYARD TENSION

Remember to distinguish between the trim of the halyard and the forestay. The forestay regulates sag and determines whether luff curve is stretched out or allowed to become fuller. The halyard regulates luff tension and determines whether the deepest point in the sail is pulled forward in the sail or dropped further aft. During practical trimming they often go hand in hand: When you need more power in the sail, you need less forestay tension as well as luff tension. When you need to reduce side forces, tightening both is a good idea. Trimming the halyard and backstay is a joint operation. Even small adjustments sailing upwind will have quite a big impact.

CUNNINGHAM: AN INVERTED HALYARD

Some boats are equipped with a cunningham in the headsail. It works just like the halyard, only now you are pulling in the opposite end of the luff close to the tack. Any difference in effect will be caused by friction in the headfoil or jib hanks.

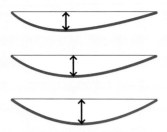

The more halyard tension, the flatter the sail. At the same time the deepest point will gradually move forwards.

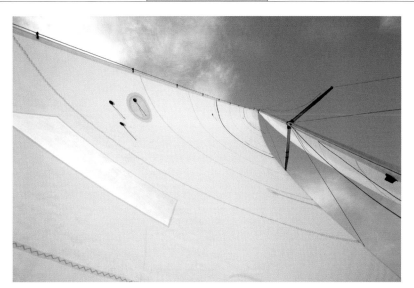

Halyard set with the right tension for the conditions (light/medium wind).

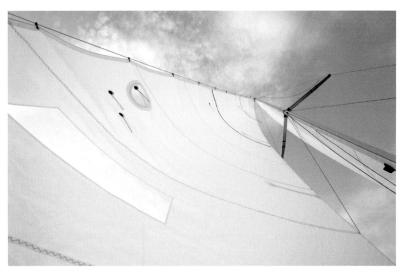

Loose halyard. Good for very light air, but here exaggerated.

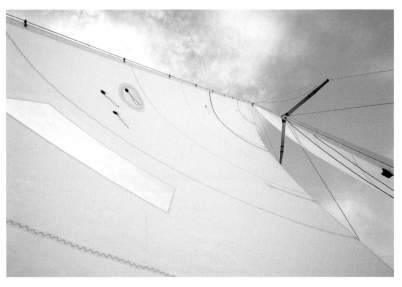

Tight halyard. Good for heavy wind, but too tight for conditions here.

SHEETING POINT

The headsail sheeting point is adjusted by moving the jib car forward or backward on the track. The sheeting point affects the twist in the sail and the depth towards the foot.

It is easy to imagine what will happen to the headsail profile if you change the angle at which the sheet is pulling. The sail is only fixed at the forestay. Both the foot and the leech are free. Pull a little more backward and it will get tighter at the foot. Pull a little further down and it will get tighter at the leech.

DETERMINES TWIST

Decreasing the sheeting angle (by moving the jib car forward on the track) will cause the sail to close more in the leech and the twist will decrease, while at the same time making the sail deeper at the foot. If you increase the sheeting angle (by moving the jib car aft) the sail will open more in the leech and twist more, while achieving a flatter foot.

FIXED POINT FOR EACH SAIL?

Sailing upwind the sheeting point is very sensitive. Even small changes will affect the boat's performance quite significantly.

Many define a good sheeting point for each headsail and stick to it, no matter the wind conditions. This may be a useful solution. Others adapt the sheeting point to the circumstances, with small adjustments on the track. The most important thing is to find the point which fits exactly to each particular sail on each particular rig. The closer the clew is to the jib car, the more sensitive the sail will be to small changes in the sheeting point.

TWIST

The main function of the sheeting point is therefore to regulate headsail twist. When the twist is just right, the sheeting point is set correctly. As pointed out earlier, when sailing upwind it is a good idea to reference the distance between the spreaders and the headsail.

USE THE TELLTALES

Telltales in the luff show if the angle of attack of the sail is good, but it is a good idea to have several telltales at varying heights. This way you can also keep an eye on the sail to make sure it is trimmed and pulling all the way up to

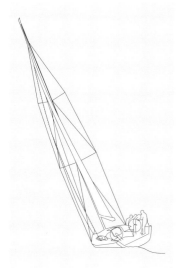

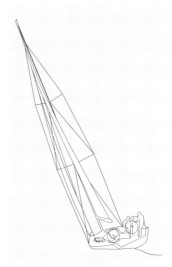

Sheeting point set right. The sail is twisting as it should. Angle of attack is vertically constant and the slot between mainsail and headsail is more or less even. This trim balances the need for power and pointing ability.

Sheeting point set too far aft. The sail is twisting too much. This will cost you both speed and pointing ability. Increased weather helm is likely (depending on mainsail trim). If combined with a well-trimmed mainsail this could work well in heavy weather. Lateral force is reduced because of the flat foot and high twist.

Sheeting point set too far forward. The sail is not twisting enough and the lower section of the sail becomes too full. Sailing like this is never a good idea. The boat will heel more and sail slower. Pointing ability is also lost.

the top. If this is the case, the telltales should react equally. If the boat starts luffing up slightly, all the telltales on the windward side should react simultaneously. If this is not the case, then it means the sail either has too much or too little twist. Either way, it is all effected by the sheeting point.

If the windward telltale at the bottom of the headsail reacts first, it means that the sheeting point is too far forward. The sail is too deep in the bottom and doesn't twist enough. If the windward telltale at the top of the sail reacts first, it means that the sheeting point is set too far aft. The sail is flat at the bottom and has too much twist. This way of controlling the sheeting point only works upwind.

SHEETING POINT ON A REACH

On open angles (when reaching) the sheeting point should be forward. When the sheet is released, the headsail usually twists more than is desirable. You can close the leech by moving the sheeting point forward. At the same time, the sail becomes deeper towards the bottom, which is also what you want to happen. On a reach the jib car can be moved quite a long way forward and it is normal to make quite large adjustments here.

However, it will be far from perfect, especially on deep courses. Even though the sheeting point is moved forward, the headsail will constantly twist too much and close at the bottom. The reason for this is that sheeting point cannot be moved far enough out from the boat's center line. A barberhaul helps, but the boat is rarely wide enough for this to solve the problem. Multihull boats have a big advantage at this point as they can move the headsail sheeting point several metres from the center line, using a barberhaul. The headsail can therefore be trimmed quite well even when broad reaching.

On monohulls this remains a problem and the deeper the course, the greater the problem. The top of the sail will usually twist out completely when the bottom of the sail is sheeted correctly.

BARBERHAUL/ TWEAKER

Barberhaul, tweaker, inhaul... all different names for a line rigged to move the headsail sheeting point sideways. You can think of it as a kind of traveller for the headsail. It makes it possible to sheet the sail closer or further from the boat's center line.

This trim feature is usually found on racing yachts and is rarely fitted on cruising boats. Even class boats for racing are often rigged without a barberhaul.

However, it is a really smart feature! The idea is simply to move the headsail sheeting point crosswise as well as forward and back, either towards the railing (to leeward) or towards the boat's centre line (to windward). You can think of it as a kind of "headsail traveller".

SHEETING ANGLE IN ALL DIRECTIONS

The genoa track moves the sheeting point longitudinally, while the barberhaul moves it laterally. Together the two trim functions make it possible to regulate the angle of the jib sheets in all directions, of course limited by the beam of the boat.

THE SUPERSTRUCTURE IS TOO WIDE

On many boats the need for this arises mostly because the jib sheet track has to be mounted on deck. The width of the boat's superstructure determines how close to the centre line the track can be mounted. Designers of cruising yachts will prioritise space below decks over upwind performance and the consequence is that on many boats the angle between the chord of the sail and the center line of the boat becomes larger than it should be for some conditions. A narrow sheeting angle will be useful upwind in light/medium conditions and especially on flat water. In waves it may

be necessary to sheet the sail further out from the center line in order to generate enough power to get through the waves with sufficient speed.

CLOSER MEANS HIGHER

Sailing upwind, the angle between the chord of the sail and the center line of the boat will determine how close to the wind the boat is able to point. The closer to the center line you can sheet the sail, the higher to windward the boat can potentially point. There is obviously a limit in the real world: when the angle gets too small you will lose speed and ultimately pointing ability as well, but on flat water, especially in light or moderate conditions, resistance is minimal. In these conditions you can optimize pointing upwind by moving the sheeting point a bit closer to the center line, while maintaining boat speed. In this situation an inhaul is very useful.

MAINTAIN PROFILE ON A CLOSE REACH

Doing, the opposite – moving the sheeting point further away from the centerline – can also prove useful. Primarily on a close reach (especially a very close reach) when you want to keep the twist more or less unchanged, while maintaining a relatively flat profile in the sail, the barberhaul can help. Without the barberhaul you would have to trim the sheet instead and possibly move the jib car to adjust twist. This would make the sail fuller and it would be likely to twist more than it should.

RIG A TEMPORARY BARBERHAUL

It is easy to rig a temporary barberhaul using a "snatch block", i.e. a block where the housing can be opened and closed. Put it around the jib sheet, in front of the car. A line running from the snatch block is passed through a block fastened out board, for instance to the toe-rail. From there it runs aft to an available winch. An inhaul/barberhaul used to pull inward toward the center line of the boat will require a very solid mounting point on the superstructure. There will be major forces at work! An inhaul is not so easy to rig, if the boat is not built or set up for such a solution.

The superstructure sets limits to how close to the center line the genoa track can be mounted. The barberhaul can pull the sheeting point further in or out.

MAINSAIL TRIM FUNCTIONS

The main is used on all wind angles and has more trim functions than any other sail. On the following pages we will look at how to work with each of them to achieve an effective mainsail trim.

HALYARD

Actually it is quite simple. It works just like the halyard for the headsail: More wind, tighter halyard. Less wind, looser halyard. Tighten the halyard upwind and loosen it downwind.

The halyard is not only used for getting the sail up and down. It has an important trim function as well. The halyard stretches the mainsails luff, moving the sail's deepest point forward. The sail gets a rounder entrance and a flatter profile, at the same time opening the leech.

TENSION HALYARD IN STRONG WINDS

Upwind, in windy conditions the halyard tension needs to be very high – higher than most people think. Tightening the halyard is a very effective way of reducing lateral forces on the boat.

LOOSEN HALYARD IN LIGHT WIND

In light wind, loosening the halyard is equally efficient. The sail will have a flatter entry and this makes it possible to sheet the headsail a bit further towards the centre line, without disrupting airflow over the leeward side of the mainsail. This improves pointing. At the same time, the sail will become fuller, providing more power and speed.

PAY ATTENTION TO MAST-BEND

Halyard and mast bend should follow each other: The more the mast is bent, the harder the halyard should normally be tightened. If you straighten the mast up by letting out on the backstay, the halyard should as a rule, be released too. This is particularly relevant in manoeuvres like bearing off downwind or heading up on a beat. The halyard should be significantly tighter upwind than downwind.

LET OFF POWER WHEN TIGHTENING

It is better to ease the sheet and take the pressure off the main before tightening the halyard. Otherwise you have to overcome a lot of resistance and there is a risk of overloading something. To establish if the halyard has the right tension for the conditions, the sheet has to be trimmed back in and the boat must be on the right heading. Only then are you able to effectively visually check the profile of the sail.

FOUR MAINSAIL QUESTIONS

1   Does the boom have the exact right angle to the wind (apparent wind angle)?

2   Does the sail open or twist as it should?

3   Is mast bend just right for the conditions – and for the design of the sail?

4   Is luff tension adjusted until the deepest point of the sail is between 40 and 50% from the luff?

CUNNINGHAM

The cunningham and halyard are quite literally two sides of the same coin – they pull at either end of the luff. The idea was originally developed to ensure that the top of the mainsail on racing boats stayed exactly on the measuring mark at the masthead at all times. This could be ensured if you pulled on the opposite end of the sail, i.e. the bottom of the luff (close to the tack) when adjusting luff tension.

If there is a lot of resistance or friction in the mast track, the halyard will pull more in the top section of the sail and the cunningham more at the bottom. A cunningham is a more practical trim tool if you want to work a lot with it as you typically would in a race. The cunningham can be adjusted

without taking the pressure off the mainsail first. This can most often not be done with the halyard.

New rigs for sportsboats are sometimes designed to de-power primarily by using the cunningham. High cunningham tension in these rigs will flatten the sail and open the leech very efficiently.

Cruising boats are rarely equipped with a cunningham, all adjustments to luff tension are made through the halyard. Dinghies and some smaller keelboats can't adjust the halyard and have to rely on the cunningham. In all cases the function is basically the same.

OUTHAUL

The outhaul is pretty straightforward: If you tighten it, it flattens the foot of the sail. Generally speaking, the lower third of the sail is affected. The more wind there is, the tighter the outhaul should be. You generally want a flatter sail upwind, whereas downwind you should let out on the outhaul to provide maximum sail power. Dead downwind (where the sail stalls anyway), it may be more efficient to keep it pretty tight, to increase the projected area. Here there are different schools of thought, but the consequences are less significant either way.

It might be tempting to sail with a very loose outhaul in very light winds even upwind, but a weak airflow stalls easily in a deep sail. A flatter profile may be more effective, even in light winds.

The trend in modern sail design is sails that are set up to be trimmed rather flat at the bottom. It is seldom effective to go upwind with a mainsail with a full foot. To some degree, the airflow will equalise under the boom, creating drag and losing much of the lift that would otherwise be created here. A full bottom section in the main will also reduce the gap between headsail and mainsail and the airflow in the lee of the mainsail will easier become disrupted.

When beating in choppy conditions in light to moderate wind, you could try letting out on the outhaul a bit and trim for a slightly lower course. The bow will pound less in the waves and the additional speed will result in less drift, which translates to better pointing. Heavy, under-rigged boats will benefit greater from such trimming.

KICK OR BOOMVANG

The kick (also known as boomvang or "kicking strap") will take over control of the twist in the mainsail as soon as you bear off and ease out on the sheet. Small boats without a backstay may need the kick to windward as well.

The boomvang or kick is an important trim function for planing or sporting boats, especially dinghies. In moderate displacement type keelboats, the kick is an important aid when sailing at deep angles in windy conditions, but otherwise it is rarely used actively on these boats.

WHEN THE SHEET DOES NOT PULL DOWNWARDS

When the boom is sheeted out going downwind, the sheet will no longer press the boom down and control twist in the mainsail. If you imagine there was no kick at all, wind pressure in the leech would pull the boom upwards. In a seaway the boom would begin to swing up and down, uncontrolled.

The kick will control the boom's vertical action, as soon as the sheet is let out.

Considering the major forces at work, the angle of the kick and it's positioning at the forefront of the boom gives it poor mechanical advantage. The kick therefore, has to be well dimensioned and high quality.

CONTROLS TWIST DOWNWIND

Initially the kicker's job is quite straight forward: It should quite simply keep the boom down when going downwind, thereby regulating the twist, or the tension of the leech. On open angles, keep an eye on the leech and consider the twist in the mainsail. If this needs to be reduced, tighten the kick. If it needs to be increased, release the kick.

The leech should be allowed to keep certain dynamics i.e. the sail must not be closed and lack "energy". At the same time, the twist should be controlled and limited. Telltales in the leech should fly all the time. If the telltales starts stalling (hide behind the sail) the kick is too tight.

BROACH CONTROL

When going downwind or when broad reaching in a lot of wind, it is important that the sail is not allowed to twist too much.

Downwind, take care to prevent the leech from being pressed forward and to leeward. This can trigger heeling to windward or rolling and ultimately result in a broach (when the forces in the sails override the helm). In heavy weather, uncontrolled twist on a dead run can cause this to happen.

The opposite is the case when sailing at sharp angles, especially with the spinnaker or gennaker set. On this point of sail, take care to allow quite a lot of twist in the mainsail. This will provide better balance and less heeling. In most cases (especially in a blow) the kick should be relatively loose on a close reach.

If you feel that you are losing control of the helm when reaching it's a good idea to drop the kick. This will increase twist in the mainsail instantly and move the center of effort forward. The boat will heel less and better balance will be achieved. Quick action here may be enough to bring the boat back on "track", allowing you to avoid a broach (where the boat is forced uncontrollably up into the wind).

BOATS WITH NO BACKSTAY

When sailing upwind the mainsheet in close cooperation with the traveller usually does the job of controlling twist. In other words, the kick will be passive.

There is however one exception. Boats without a backstay – dinghies, fast multihulls and planing keelboats – will also often benefit from using a kick upwind, especially in a blow. Upwind, the main sheet is worked actively in gusts and waves, especially in these boat designs. The kick will help to maintain mast bend and forestay tension, even when sheeting out. You do not want to lose longitudinal rig tension every time you release the sheet. Assuming the rig is dynamic and not too stiff, the kick will handle this job very well. Trimmed like this, you will not lose the flattened, open profile of the sails whenever you sheet out in on a gust. Twist will of course also be under control.

The kick is adjusted until the mainsail twists more or less as it should when reaching or running.

TRAVELLER

The traveller makes it possible to move the mainsail sheeting point sideways. This means you can change the sail's angle of attack without changing the twist – or visa versa. These options are useful in many situations.

CONTROL OF THE ANGLE OF ATTACK AND TWIST

When trimming the main upwind to maximise pointing ability, it is important that the boom is more or less on the boat's center line (in some boats slightly to leeward). This is the boom's "power position" and it should be left here until side forces makes it necessary to reduce the power of the mainsail and control heeling and weather helm. Only then should the boom be moved to leeward. If you need some extra pointing ability, you can even try to pull the boom slightly across the center line, to windward. This will in most cases cost you a little speed, but can still be a useful tactical tool in a race situation. Twist is another major concern. The traveller makes it possible to adjust sheeting point and sheeting until the twist is just the way you want it to be, while keeping the boom exactly where you want it. The traveller and sheet combination gives full control over the angle of attack and the twist at the same time.

Traveller to windward. Sheet eased to achieve a lot of twist. Good light air trim.

Traveller to leeward. Sheet eased. The boom is further out, but the amount of twist is the same as above. Good heavy weather trim and well suited for very close reaching.

Traveller in the middle. The boom is at the same angle as in the top illustration. Tight sheet leads to tight leech (reduced twist). The back of the sail is stalling. This is a frequent trim mistake and costs speed.

ADJUST THE TWIST?

If you want more twist while keeping the angle of attack unchanged, move the traveller to windward and ease the sheet. The leech will open and the boom will stay in the same place.

If you want less twist, while keeping the angle of attack unchanged, move the traveller to leeward and tighten the sheet. The leech will close and the boom will stay in the same place. This is typically something you would play with in light and variable wind conditions, when you need to adapt twist with small increments on a beat.

The traveller is trimmed to leeward to reduce side force in strong winds.

ADJUST ANGLE OF ATTACK?

If you want to keep the twist, but change the angle of attack, move the traveller without touching the sheet. This is typically something you would do if you need to adjust the balance (helm) or if you want to reduce heel in stronger winds. The most common way of playing this is to drop the traveller to leeward in gusts and pull it back up in lulls. This way you will make sure heeling and weather helm stays more or less constant, while maximum power is maintained. This is a fine balancing act and requires constant attention if you want the boat to sail optimally at all times.

WHY NOT JUST USE THE KICK?

The traveller has little or no effect whilst sailing downwind. As shown on page 142, the kick or boomvang is used to control twist on open wind angles. You could say that the kick and traveller are related, as both are working on twist in the mainsail, but the traveller only works upwind or at very sharp wind angles (depending on how long it is). The kick works best when the boom is sheeted far out. In principle, the kick could replace the traveller and does so in some dinghy classes. This only works on very small boats. Upwind sailing creates forces that quickly become so large, that in most boats the kick would not be powerful enough. The leverage is simply not high enough.

If kick was the only way of controlling twist, you would also miss the opportunity to give the mainsail extra twist going upwind in light air, by moving the sheeting angle to windward. This can only be done with a traveller.

WHY NOT JUST USE THE SHEET?

You may well use the sheet instead of the traveller. In fact, a lot of cruising boats do not have a traveller at all. However, as we have seen, this has its consequences. Every time you release or tighten the sheet without compensating with the traveller, you affect both the twist and angle of attack. You cannot change one without changing the other.

If you want to adjust balance or heel by letting the boom out using the sheet alone, you will increase twist and change mainsail profile as well as change the angle of attack. This will of course limit your trim options.

If you want maximum power and pointing ability upwind, you need to tighten the sheet to bring the boom to the center of the boat. To bring it all the way in without a traveller, you may be forced to tighten the sheet too much, ending up with a closed, stalling leech.

Using the mainsheet only going upwind is not necessarily a bad thing. In some circumstances it can work well. By exclusively using the sheet you will get a closing leech when the boom is brought to the middle of the boat and an opening leech when the boom is let out. That can be an acceptable way of dealing with gusts, as you will instantly release excess force from the head of the sail and at the same time move the center of effort forwards.

MAINSHEET

The mainsheet has primarily two jobs: To control the angle of attack and the twist. Upwind, both jobs are done in close cooperation with the traveller. Downwind, the kick takes over twist control and leaves only the angle of attack adjustment to the sheet.

While sailing upwind the mainsheet pulls more or less straight down from the boom. As a consequence, pulling on it will not only change the angle of attack, but just as much leech tension, which will effect the twist. As shown on page 144, the traveller and the mainsheet work together while upwind, fine-tuning the relationship between angle of attack and twist.

Downwind the mainsheet pulls in a much more horizontal angle. Here, the mainsheet will take care of the angle of attack all by itself. As soon as the sheet is let out, the kick will take over the job of controlling twist. Once you bear off on a reach or run, the traveller is out of the game.

TELLTALES IN THE LEECH
When it comes to upwind mainsheet trim, telltales are very useful tools. You can read more about telltales on page 122.

The top telltale in the leech (which should preferably sit at the tip of the upper batten), deserves the greatest attention. Finding just the right amount of twist and angle of attack for the conditions present is what upwind mainsheet trim is all about. This telltale will tell you.

80/20
If the telltale on the top batten is hiding behind the sail all the time, the leech closes up too much. The head of the sail is stalled. If the telltale flutters freely, without any tendency to hide behind the leeward side at all, the leech opens too

much. Power is lost. The ideal leech tension is where the top telltale flutters between these two positions, preferably a little less behind the leech than straight out. 80% out, 20% behind can be your starting point, but if you want to squeeze a little extra pointing ability out of the boat, it may be more along the lines of 50/50. This applies in "power" mode, in other words up to the point where side forces makes it necessary to release excess power. When this happens, more twist is a good response, but only when more halyard, outhaul and backstay have already been applied. When you are overpowered, your sheet trim should focus on other things than telltales, such as heeling and helm pressure.

TOP BATTEN PARALLEL TO THE BOOM
The top batten itself is also a good reference point for the mainsail trimmer. If your look is aiming upwards in the sail from under the boom, you can easily see if the top batten is parallel to the boom. It may either fall out to leeward or point to windward.

The goal is to trim the sheet and traveller until the top batten is parallel to the boom, with the boom still in it's power position (at the center line). In very light air or choppy seas, it may serve you well to twist the mainsail a little extra. The same goes of course for heavy weather. In medium winds, in situations where you would benefit from a bit of extra pointing ability, you could try trimming the sheet until the upper batten hooks a bit (points slightly to windward).

Too tight sheet, too little twist. Top batten "hooks" (points inward). The head of the sail is stalling.

Looser sheet, more twist. Top batten is parallel to the boom. Good trim for light/medium conditions.

Even looser sheet, even more twist. Now the boom starts moving to leeward. Good trim for heavy weather and gusts.

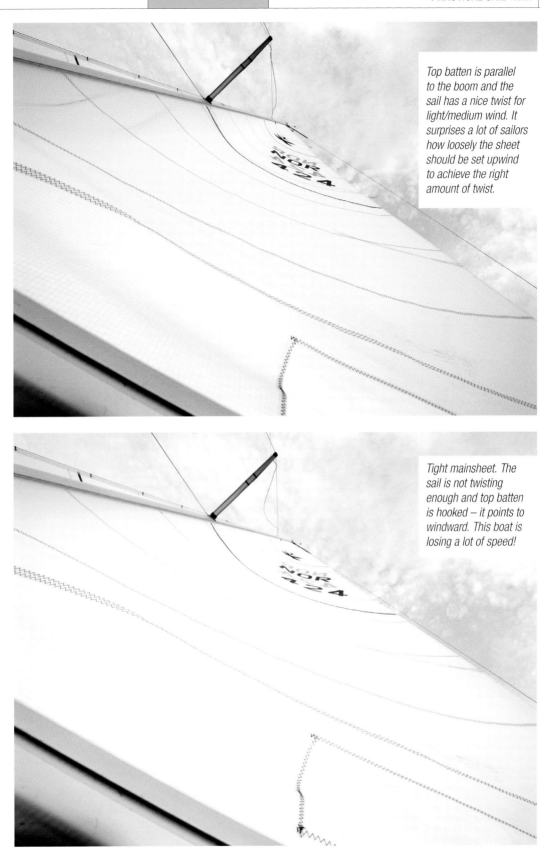

Top batten is parallel to the boom and the sail has a nice twist for light/medium wind. It surprises a lot of sailors how loosely the sheet should be set upwind to achieve the right amount of twist.

Tight mainsheet. The sail is not twisting enough and top batten is hooked – it points to windward. This boat is losing a lot of speed!

TRIMMING THE MAINSHEET ON A REACHT

The kick takes care of mainsail twist when sailing off the wind. Trimming the mainsheet now is simply about creating maximum power in the mainsail. On a broad reach it can actually be quite challenging to trim the mainsheet correctly because there are not so many reference points. A very common mistake is to trim the sheets too hard – especially in light/medium conditions. Laminar airflow on the leeward side of the sail is lost, stalling occurs and speed drops. A good tip is to let out slightly on the mainsheet regularly, to the point where the sail begins to luff, or where windward telltales start to flutter. Then tighten, just enough to correct it. Sheet pressure will tell you when you are close to maximum power. Twist can be checked by looking at the telltale on the top batten, just as you would do upwind. If in doubt, ease the sheet!

TRIMMING THE MAINSHEET DOWNWIND

Downwind all hopes of a laminar air flow and efficient lift production is in vain. The wind will simply push on the windward side and the leeward side will inevitably be turbulent, with separated airflow. In this situation, the more sail area you can expose to the wind, the better the results. Ideally, the sheet should be released so far out that the sail is situated more or less perpendicular to the boat's center line.

Note that it is the sail that should be be perpendicular to the boat, not necessarily the boom. Twist can be significant, depending of course on how hard the kick is set. In heavy weather, make sure the mainsheet and kick is a little tighter than in lighter conditions. If the leech ends up forward of the mast and the boat is sailed at a very low angle (close to a gybe), the boat will become very unstable and rolling can occur.

The spreaders (especially swept spreaders) will limit how far out you can feed the sail. Try to reduce wear from the sail pressing on the spreaders.

A much more common mistake, however, is to trim the mainsheet too tight going downwind.

MAINSAIL TRIM IN GUSTS OF WIND

There are two schools of thought here: ease the sheet or let go on the traveller. Using the sheet is possibly the best solution on boats with a backstay, but there are differing views. Using the traveller is definitely the best option on boats without a backstay.

Upwind, immediate action is crucial if you want to sail through a gust without losing speed. The boat is suddenly overpowered and excess force must be let out of the mainsail. If you don't act before the boat heels over, you will end up with more drift, more weather helm and less progress. Building speed again will take a while and in a competitive race fleet this could be costly.

MORE TWIST AFFECTS THE HEAD;2

There are two basic philosophies: Some like to keep the traveller where it is and ease the mainsheet when a gust hits. Others like to use the traveller and leave the mainsheet where it is. Let's first take a look at what happens when you ease the mainsheet in a gust. The idea is that increased twist releases excess power from the top down and as a result, side force will reduce quicker. As soon as the wind is back to normal, the sheet is trimmed back in. This technique works best on boats where the backstay (or other types of rigging) is able to keep mast bend and forestay tension unchanged, even when the mainsheet is eased.

HOW TO KEEP LONGITUDINAL RIG TENSION

On boats without a backstay or running backstays (usually dinghies or sports multihulls and planing keelboats) the mainsheet will pull on the mast through the mainsail leech. The leech can be seen as a substitute for the backstay. On these boats, the mainsheet will to some degree control longitudinal rig tension, often in combination with the kick and shroud tension. If you let out the mainsheet in a gust on these boats, mast bend and forestay tension is immediately reduced. Shroud tension and kick will help, but most often this is not enough. The result is more sag and a straighter mast. The sails will become fuller and the consequence

The sheet is eased in the gusts. On this Laser, the kick keeps the mast bent and the sail flattened when the sheet is let out. On a stayed rig, this will normally be executed by the backstay.

is more side force – not less. In a worst case scenario, easing the mainsheet could actually increase heeling, not reduce it.

With a traveller on the other hand, you can maintain your mainsheet tension and still let out the boom to reduce the mainsail's angle of attack. Forestay tension and mast bend is preserved through the gust and you are still able to re-duce heel caused by the mainsail.

THE LEECH WORKS AS A BACKSTAY

The traveller is a huge help on a boat with no backstay. You can work with the mainsail's angle of attack, rather like flaps on an airplane wing. When the boom is let out, there is obviously less power in the sail. The power still produced will be directed more forwards and less sideways. When the wind speed drops back to normal, pull the traveller back until heel angle and weather helm feels right.

This will actually work very well in any boat. But in boats rigged with a backstay choosing sheet or traveller is mostly a matter of taste. As you can see on page 124, active use of the backstay should also be a part of the mainsail trim, especially in gusty conditions. In fact, on some boats the crew will use the backstay more than any other trim func-tion to depower the rig when a gust hits.

KICK, THE LAST BASTION

Some boats have neither traveller nor backstay. In that case you have to rely on the kick (assisted by shroud tension), to maintain longitudinal rig tension when the mainsheet is let out. This requires extremely high kick tension, but executed correctly it can make a big difference in the small boats that are rigged this way.

FIVE TRIM FUNCTIONS

On the following pages, we will look at the four trim functions used for the spinnaker: Topping lift, after guy, barberhaul and spinnaker sheets. We will also look briefly at the lines used to set or stabilise the sail: halyard, foreguy and mast track.

SPINNAKER TRIM FEATURES

The spinnaker is a sensitive and unstable sail with a lot of power. A big part of spinnaker trim is about stabilising and controlling the sail. The spinnaker requires attention, but gives a lot of speed and fun in return.

The spinnaker is symmetrical. When you gybe the two sides stay in the same place, but they change their name and function. Nevertheless, it always has a luff (towards the wind) and a leech (away from the wind).

The spinnaker is a so-called flying sail – it does not have a forestay or mast track where the luff is fixed. The spinnaker must produce enough lift to fly by itself. In fact, it only has two fixed points: the top of the sail and the tack – the corner of the sail attached to the end of the spinnaker pole. The rest of the sail is flying. The spinnaker is made of flexible nylon cloth, with far more stretch than any other sail.

UNSTABLE

All in all, this makes the spinnaker a very unstable sail. It collapses easily, it can suddenly pull in a different direction than the heading of the boat and it can start swinging from side to side as well as back and forth, especially in waves. It is a large sail (usually bigger than both mainsail and headsail combined) and the forces involved can quickly become violent. Handling (hoisting, gybing and dousing) is more complicated than with other sails. All in all, a spinnaker can be a handful. For cruisers good advice is to use the spinnaker in relatively light winds (not more than 10-12 knots) and only for broad reaching or running (headings well aft of the beam). It is a good idea to take it down early if the wind starts to pick up.

SAFETY, SPEED AND STABILITY

This instability is an important element when trimming a spinnaker. One main concern is to stabilise the sail. It is not just about safety or comfort; a spinnaker that swings back and forth, perhaps even threatening boat control, is not an efficient sail. Instability is never good for a sail. A swinging spinnaker disrupts airflow and lift, preventing the sail from pulling as it should. When the spinnaker is rolling, the boat starts to roll too and the associated flow of water over the rudder, keel and underwater hull will affect steering and create drag. When trimming, first priority is to stabilise the spinnaker and avoid rolling. Safety, speed and stability go hand in hand.

GUY, TOPPING LIFT AND BARBERHAUL

The stability of a spinnaker is primarily a result of how the two lower corners of the sail are positioned: Firstly the location of the tack, both vertically and horizontally (the same as the position of the outer end of the spinnaker pole) and secondly the clew (which is flying and dependent on the sheeting and the sheeting angle).

The tack is controlled by the guy, the topping lift and the downhaul. The clew is controlled by the sheet and the barberhaul, if there is one.

CRUISING? LET THE HELMSMAN DO THE JOB

If you want to sail optimally with a spinnaker, it requires active trimming. This is one of the reasons why spinnaker sailing is not that popular among cruisers. Cruisers can however enjoy a leisurely voyage in a nice breeze with a spinnaker up. If the guy and topping lift are set correctly and the sheet is over-trimmed a bit, the spinnaker can be set up with a steady base trim adapted to the current wind angle. As soon as that is done the crew can relax and leave it up to the helmsman to steer by the curling luff.

Trimming the spinnaker sheet is really very simple (at least in theory!). It should simply be trimmed as loose as possible at all times.

Actually, a lot of sailors find spinnaker sailing fun. It can activate the entire crew or family, making it a great sail for cruising as well, but once again – only when conditions are right. When the wind is up or the wind angle is tight, the spinnaker should be treated with respect.

CONCEPTS, EQUIPMENT AND HANDLING

As you will see on the following pages, the spinnaker is a bit more complicated to use than other sails. Sailing with a spinnaker requires the boat to be rigged with special equipment and there are some manoeuvres you should know in order to set, gybe and douse the spinnaker. In this book we have chosen to omit descriptions of manoeuvres and maintain focus on sail trim. There are other books on the market that focus on the practical handling of spinnakers.

THE THREE CORNERS

Let's look briefly at the three corners of the spinnaker. As mentioned earlier, there are two fixed points (head and tack) and one flying point (clew).

The head of the sail sits right over the top of the forestay and there is not much to be done with it. It is quite simply a fixed point. The head is hoisted to the top and with few exceptions it remains there, meaning that the spinnaker halyard is not regarded as a primary trim tool.

THE TACK

The second fixed corner on a spinnaker is the tack, attached to the end of the spinnaker pole. The tack can definitely be trimmed. The spinnaker pole can be moved vertically and horizontally. The afterguy moves the tack in a vertical direction and the topping lift moves it in a horizontal direction.

The foreguy (which in some boats is replaced by a barberhaul on the afterguy) is not really a trim tool, it simply serves to keep the spinnaker from pulling the spinnaker pole upwards and stabilises the tack.

A lot of spinnaker trim is about positioning the tack optimally to wind angle, wind speed and (not least) wave conditions at all times. This is where you can manage the profile of the spinnaker and the luff tension. This is also where the angle of attack is determined (in collaboration with the sheet).

THE CLEW

The third corner is the clew, which is flying. In addition to tightening and loosening the sheet, the sheeting angle can normally be adjusted too, just like the with headsail. The sheeting block usually sits at the back of the boat, but by using a barberhaul you can move the sheeting point forward, changing the sheeting angle. This will reduce twist and stabilise the spinnaker. Not all boats are equipped with a barberhaul, but this can be a useful trim tool, especially downwind when the wind is up.

The tack (left) is pulled close to the spinnaker pole.
The tack can be moved, but is stabilised with afterguy,
spinnaker pole, downhaul and topping lift. The clew (right)
is flying.

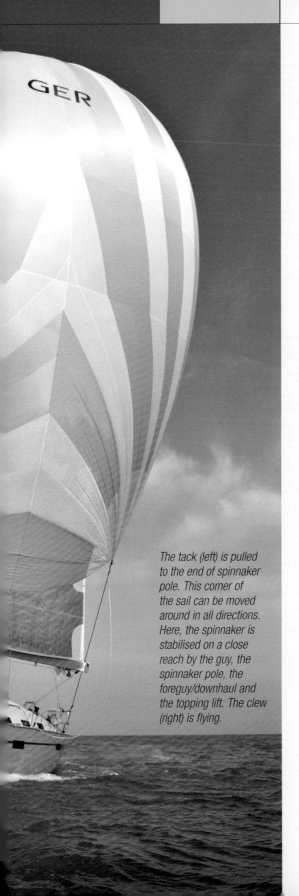

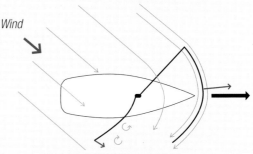

Laminar airflow in spinnaker.

A SAIL LIKE ANY OTHER SAIL

A good tip is to think of the spinnaker as a sail just like any other sail. It has an entry and an exit – in other words a luff and a leech. You need to maintain a certain amount of twist in the leech. The airflow moves (more or less) from the luff to the leech and should (as far as possible) follow both sides of the sail. At low wind angles however, a portion of the airflow will be vertical instead of horizontal as air will flow over the top of the sail and down on the leeward side.

The tack (left) is pulled to the end of spinnaker pole. This corner of the sail can be moved around in all directions. Here, the spinnaker is stabilised on a close reach by the guy, the spinnaker pole, the foreguy/downhaul and the topping lift. The clew (right) is flying.

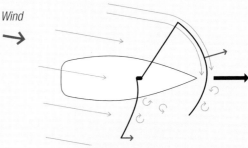

Stalling in spinnaker.

STALLS DOWNWIND

When running before the wind the spinnaker will stall (just like any other sail). When you bear down past 160-170 degrees (running) the angle of attack becomes too big and laminar airflow will separate and become turbulent. The name of the game now is to present the greatest possible surface area square to the wind. Some fullness can still provide partial laminar airflow running vertically over the top of the sail.

LINES NOT USED FOR TRIM

The spinnaker is trimmed using the afterguy, topping lift, sheet and barberhaul, but there are also a few lines attached with other functions.

MAST TRACK FOR THE SPINNAKER POLE

Smaller boats will often have a hoop or ring at the front of the mast for attaching the spinnaker pole. But on bigger boats the pole can often be moved up and down on the mast. If that is the case, the pole end (beak) sits on a bracket that runs on a vertical track mounted on the front of the mast. The idea is to make sure the pole does not lose effective length, when the other end (the tack) is being lifted or lowered as part of the spinnaker trim. A long spinnaker pole is better than a shorter one and a pole set horizontally is optimal for length. It also becomes stronger.

Minor adjustments to the topping lift, as part of ongoing fine tuning, can be carried out without compensating at the mast track position, but major trim changes on the topping lift should be followed by the spinnaker pole being moved up or down on the mast track until the pole is horizontal.

FOREGUY/DOWNHAUL

The spinnaker will try to pull the spinnaker pole up in the air as soon as the wind is strong enough. You need to be in charge of this, since it affects the profile and stability of the sail. This is why most boats are equipped with a spinnaker pole foreguy.

On smaller boats this job is often left to the afterguy. In reality the afterguy barberhaul will press the spinnaker pole down. Larger boats require more power and must be rigged with a special downhaul or foreguy for the purpose.

Some boats have the foreguy attached to the foot of the mast. The spinnaker pole can swing around without the need for adjustment, which makes it easier to handle. But on boats where there are major forces at play, the foreguy needs better leverage and has to run straight down to a block on the foredeck. That makes it necessary to adjust the foreguy whenever the afterguy is trimmed. However, this solution gives better stability to the spinnaker pole, as it also prevents it from swinging back and forth in light winds and waves.

HALYARD

In contrast to the halyard in the other sails, the spinnaker halyard does not have an important trim function. Usually you simply hoist the sail right to the top and leave it there. Some sailors like to drop it a little to help the spinnaker further off the mainsail on a run, but this is rarely effective. As a rule, the halyard is hoisted all the way to the top.

SPINNAKER TOPPING LIFT

The topping lift or spinnaker pole uphaul does what the name implies. With this line you set how high or low the tack will sit. The foreguy/downhaul will stabilise it. This is where we trim luff tension, roughly the same way as with a cunningham.

An old rule of thumb says that both corners of a spinnaker should fly horizontally. This can be a starting point (at least downwind) but it is not necessarily correct. Quite often the tack should be set lower than the clew. The tack should ideally never be higher than the clew, but in very light winds this can be difficult to achieve as the weight of the sheet pulls the clew down and closes the leech. Lightweight sheets for light airs are a good idea!

LOW SPINNAKER POLE WILL STABILISE
If you lower the spinnaker pole you will increase luff tension and stabilise the spinnaker. At the same time the clew will be lifted, which will increase twist. In very light air, when the sail is about to collapse, this is helpful. The spinnaker will also be able to fly easier. The same applies in choppy waves, when the spinnaker swings from side to side and may even collapse due to the movement of the rig.

HIGH SPINNAKER POLE, MORE POWER
In light/moderate to windy conditions you should normally raise the spinnaker pole, increasing it as the wind angle becomes deeper. The head of the spinnaker will be wider and flatter and the increased presented area will give you more power and more speed, especially on a run. When the wind increases to a point where there is a risk of broaching or when the crew struggles to maintain control, the focus shifts. Now you should trim for stability and control. In this situation, it helps to lower the pole a little and tighten the

barberhaul on the spinnaker sheet. The spinnaker will twist less in both the luff and leech and with the sheeting point further forward it will be more stable. Heeling can trigger a broach, so it is important to keep the boat as flat as possible.

TOPPING LIFT AFFECTS THE MAST TRACK, FOREGUY AND AFTERGUY
When trimming the topping lift you also have to adjust the other lines that affect the spinnaker pole. Exactly what has to be done varies slightly depending on how your boat is rigged, but here are the most common things:

- The foreguy/downhaul must be adjusted accordingly, to fix the position of the spinnaker pole.

- The other end of the spinnaker pole must be moved up or down on the mast track, to keep it horizontal (at least when more fine-tuned).

- It may not be necessary to adjust the afterguy, but the horizontal position of the spinnaker pole is also affected by the topping lift – especially when the boat is rigged with a barberhaul. Raising the spinnaker pole without adjusting the afterguy will move the pole slightly backwards. Lowering the spinnaker pole without adjusting the afterguy will move it slightly forward.

1.
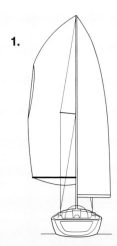

2.
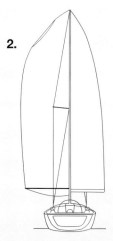

3.
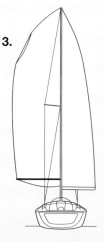

LOW SPINNAKER POLE

When you lower the spinnaker pole, you increase luff tension. As in any other sail, this will move the deepest point of the sail in the windward direction, in this case closer to the pole. The entry will become rounder and more "forgiving", the leech will open (higher clew, more twist on the leeward side) and the spinnaker head will become deeper and more closed. Also the shoulder will narrow. All in all these are good characteristics for reaching. The higher the wind angle, the lower the pole. Here it is exaggerated. The clew flies very high. The leech opens too much and power is lost.

HIGH SPINNAKER POLE

When you raise the spinnaker pole, you decrease luff tension. The deepest point of the sail will move toward the center. The entry will become flatter and the sail will be more sensitive to trim and steering. The leech will close (lower clew, less twist on the leeward side) and the spinnaker head will be become flatter and wider. The shoulder gets broader. These are good characteristics for deep wind angles. Exaggerated, (as here) the clew ends up lower than the tack. This is never a good idea.

WHERE DOES IT BREAK?

When the luff aerodynamics start to break or cause it to curl, it tells us that the sheet is trimmed correctly. But the position of the curl can also tell you if the topping lift is trimmed correctly.

1   If the spinnaker pole is set too high, luffing will occur near the bottom of the sail.

2   If the spinnaker pole is set too low, luffing will occur near the top of the sail.

3   When the topping lift is trimmed correctly, luffing will occur about two-thirds up and be distributed over a larger area. The sail will not be as prone to collapse as soon as the luff curls.

AFTERGUY

The afterguy sits at the end of the pole and leads the spinnaker around the mast in an arc. This controls the angle of attack, in close cooperation with the spinnaker sheet. The afterguy should be pulled backwards as wind angles get deeper.

The afterguy is the spinnaker's windward "sheet" – the line running through the spinnaker pole. When gybing, the afterguy and spinnaker sheet switch names. Sheeting as such only happens at the clew on the leeward side, so it is appropriate to refer to the afterguy with a different name. Larger boats can be equipped with double sheets and double guys, i.e. two lines attached to each of the two lower corners of the sail. The purpose is to simplify the somewhat tricky process of gybing a spinnaker.

BARBERHAUL

The boat should be rigged with a barberhaul on both sides and the barberhaul on the windward side (associated with the afterguy) should always be pulled all the way down. When trimmed like this, the afterguy will to some extent pull the spinnaker pole down, in conjunction with the foreguy (if rigged).

DETERMINING THE SAIL'S ANGLE OF ATTACK

The main function of the afterguy is to decide the position of the spinnaker relative to the boat. You can imagine the spinnaker being led around the mast in an arc at the end of the spinnaker pole, presenting the luff to the wind. In other words, the afterguy (and the following sheet) will determine the spinnaker's angle of attack. Again, we are trying to achieve maximum laminar airflow.

The afterguy will also trim the fullness of the lower sections of the spinnaker. Ease the afterguy forward and the spinnaker becomes deeper. Pull the afterguy back and it becomes flatter.

The rule of thumb is that the spinnaker pole should be set 90 degrees to the apparent wind angle. But as with rules of thumb in general, this is not always true.

FORWARD ON A REACH, PULLED AFT DOWNWIND

In principle you ease the afterguy and the spinnaker pole towards the forestay as you sail higher to the wind and pull the afterguy and spinnaker pole back towards the shrouds when you bear away.

Pull the spinnaker well out to windward at deeper wind angles. This will bring the spinnaker out of mainsail wind shadow and improve balance, as a portion of the spinnaker now flies on the windward side of the boat.

AFTERGUY DOWNWIND

Downwind, when the sail stalls, bring the pole back and out to present the largest possible sail area to the wind. This is a balancing act, however: a deep spinnaker allowed to fly high may have a smaller presented area, but in return it will provide more laminar airflow at the head, along the luff and possibly even the leech. A flatter spinnaker, pulled down with the foreguy and back with the afterguy has a larger area, but also a more separated airflow. A balance between these two extremes is what you should aim for.

In heavy weather you can stabilise the sail and the boat by pulling the spinnaker pole down and carry the pole a little farther forward than normal. When trimmed like this barberhaul on both sides should be tight.

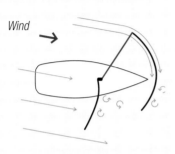

Wind

Tight afterguy on a run.

VERTICAL LUFF

Generally, the afterguy is trimmed until the luff (shoulder) has a smooth, rounded arch – more so at lower wind angles. The shoulder should not point inward toward the boat's center line or outward, away from the boat. The shoulder located vertically above the end of the spinnaker pole is a good indication. The lower part of the cap shroud (below the first spreader) is a useful reference. The luff should be more or less parallel with the shroud.

From left:
1. *Too tight afterguy*

2. *Correctly trimmed afterguy*

3. *Too loosely set afterguy*

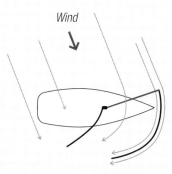

Wind

Eased afterguy on a reach.

AFTERGUY ON A REACH

To make spinnaker sailing on sharper wind angles possible, the spinnaker has to be eased in an arc with the wind (to leeward), maintaining the angle of attack as the boat heads closer to the wind. Lateral forces will increase and the boat will heel more and this is most often the limitation for how close to the wind you can go before you have to come off the wind or douse the spinnaker.

Keep some space between the spinnaker pole and the forestay. In gusts the spinnaker pole will be pushed hard against the forestay and this can damage the forestay and the pole.

Remember that increased lateral force also means decreased forward thrust. If pushing too hard to windward you risk powering the boat to the limit, but most of the power will heel the boat, increasing drift. In this situation you would be better off with a headsail. Logically you can sail significantly closer to the wind in lighter airs. When reaching in strong wind, it helps to lower the pole. This gives the spinnaker a "gennaker profile". The clew will fly higher and further from the boat (barberhaul off). Leech twist will reduce side force.

HELMING AND SHEETING

Spinnaker sailing involves active steering and trimming of the sheet. But how is it done – and why does it work?

Sheets are always first priority when trimming sails. This applies to the spinnaker as well. If you want to sail optimally with a spinnaker, one crew member has to be fully focused on the spinnaker sheet. His or her mission is simply to make sure that the luff is more or less constantly curling – constantly on the edge of collapse. This indicates laminar airflow at the luff (the entry of the sail). In other words, the sheet should be eased out as far as possible.

GUST: WIND MORE ASTERN

It is easy to get the impression that gusts will cause the apparent wind angle to become sharper. However, with displacement boats the opposite actually happens. The boat is probably already sailing at maximum hull speed, so when true wind speed increases, this increase is not matched by an increased induced wind. Boat speed is therefore more or less constant. Induced wind becomes a smaller component and apparent wind angle becomes broader. To sum up: when a gust hits, apparent wind will shift to a broader wind angle. This is why the trimmer needs to ease the sheet quickly when a gust hits and trim back in when the wind speed drops (or, if boat speed actually does increase).

SPEED INCREASE: WIND MORE FORWARD

In light wind, boat speed is unstable and in gusty conditions the pattern will appear a little messy. Once you know the mechanisms however, it is logical enough. Ease the sheet quickly when the gust hits, tighten it as the boat speed picks up. When wind speed drops, boat speed will still be high, so for a few seconds you will have to sheet in even more, because induced wind now becomes a larger component. Seconds later, the boat speed will drop and the sheet must suddenly be let out – and quite a lot!

Understanding this makes spinnaker trimming easier, but strictly speaking you can get along very well by paying close attention to the behaviour of the sail. Ease out until the luff curl appears and tighten slightly until it is almost gone. During racing this process is continuous, with adjustments every few seconds.

ACTIVE HELMING

Helming off the wind will of course affect the sail's angle of attack, just as much as it does when sailing upwind. Changes in the apparent wind angle may be met by changing course, instead of changing trim. This is not only relevant in racing situations: Active helming can simplify spinnaker sailing and help facilitate it's adaptation to cruising. In stable, relatively calm conditions you can lock the sheet and control the luff with minor steering adjustments.

GO DEEP IN GUSTS!

Whatever your objective and level of ambition is, it makes sense to steer actively when sailing with a spinnaker. It provides greater security and control, as well as better speed. A simple rule of thumb is to come off the wind whenever power increases and luff whenever power decreases. In very light air this is a very useful way to keep the spinnaker from collapsing. Of course you need an active trimmer to make it work well.

AVOID A DEAD RUN IN STRONG WINDS

In a blow you need a safety margin from a dead run (180 degrees). 15-20 degrees above this gives you the option to bear off if you need to, without risking an unexpected gybe. In heavy weather, try to avoid dead downwind sailing altogether. The spinnaker can easily start to pull to windward and force the boat into a broach. A chinese gybe, as this is called, is the worst form of involuntary broach and when using a spinnaker in windy conditions, you should make it a first priority to avoid this happening.

STAY IN THE GROOVE

But how is that done? You will of course need the right heavy weather trim for the foreguy, afterguy and barberhaul, as described on page 152 to 161. But active helming and sheeting is what keeps the boat upright through the gusts. Check out the tips on the next page.

There is a fairly narrow "groove", point or lane, where the boat is relatively stable. This "safe lane" varies slightly from boat to boat, but is usually around 150-160 degrees (apparent wind angle). Here the wind angle is broad and lateral forces are reduced, but there is still a good safe margin before a dead run appears and the associated instability and risk of a chinese gybe that lurks as you get close to 180 degrees. Most broaches start with heeling or rolling. Try to stay in the groove, sail the boat flat and stabilise the spinnaker.

4 TIPS FOR HEAVY WEATHER

➕ Steer towards the direction of the masthead when the boat starts to roll. The bow should be under the spinnaker! This will stabilise the boat and keep it flat.

➕ Watch the windex or wind instrument. Know your wind angle! 150-160 degrees apparent wind angle will often give you a stable heading.

➕ Pay attention to weather or lee helm. If pressure increases, deal with it immediately. Is the boat forced into the wind? Come down quickly and ease out on main and spinnaker sheets (ease kick, if possible). Is the boat forced down with the wind? Come up quickly and tighten the spinnaker sheet (ease afterguy if possible).

➕ "Hourglass" is when the spinnaker twists and ties itself and possibly even ends up wound around the forestay. It happens mostly while hoisting or gybing. The reason is mostly turbulence from the mainsail. Often the hourglass will resolve itself if you gybe the mainsail. The reversed turbulence is likely to twirl the spinnaker back and loosen the hourglass. If not – take the sail down.

MORE TIME IN THE FAST LANE

Safety and control are not the only reasons why it is a good idea to go low in gusts and come up between them. This is actually an advantage in all conditions and will actually result in greater average speed overall. The reason can briefly be explained as follows:

All sailing boats have a particular wind angle where they achieve maximum hull speed. This wind angle is broad in strong winds and gets narrower as wind speed drops.

By luffing whenever the wind speed drops and bearing away whenever the wind picks up, the boat will ultimately spend more time on a faster wind angle. This benefit is very significant in planing boats and more modest in displacement boats, but all boat types will take advantage of such tactics. See more on VMG sailing on the next page.

WHAT IS A GENNAKER?

A gennaker or asymmetric spinnaker is a kind of cross between a spinnaker and a genoa – hence the word "gennaker."

The basic shape is from the genoa: a defined tack and luff in front and a leech in the back with both sheets in the clew and an asymmetrical shape (the luff is longer than the leech).

From the spinnaker it has the flying luff and a big, curvy shape. The cloth is also the same: light and flexible nylon spinnaker cloth.

TRIM FEATURES OF THE GENNAKER

The gennaker is considerably easier to manage than the spinnaker and is trimmed primarily by the sheet. On a reach the two sails share many of the same characteristics and they are trimmed similarly. But there are also differences.

A significant characteristic of the gennaker is that the tack is attached to a line led directly to the bow – or to a bowsprit. Unlike the spinnaker with its afterguy and spinnaker pole, you cannot move the tack of a gennaker horizontally and adapt it to different wind angles.

The tack of the gennaker is a fixed point and remains more or less on the center line of the boat. When you bear off to a deep wind angle, the gennaker will at a certain point be affected by the mainsail wind shadow.

EASY TO HANDLE

It is possible to set a gennaker on a spinnaker pole. But one of the great benefits of the gennaker – its simplicity – will be lost. Basically the gennaker is a very simple sail, with few trim functions. Compared to a spinnaker it is also much easier to hoist, gybe and take down.

Cruisers will typically want an easy manageable form of spinnaker, without the troublesome spinnaker pole. The gennaker can be handled shorthanded and requires less special equipment. This is the main reason for its popularity among cruisers.

SHARP ANGLES AND FAST BOATS

The gennaker also attracts an entirely different group of sailors. Lightweight boats with a lot of sail and planing capabilities (dinghies, multihulls, or keelboats specifically designed to plane) will also benefit greatly from this sail and for entirely different reasons.

Gennakers can be used on closer wind angles than spinnakers. They may not have the same versatility and power when it comes to deep wind angles, but they can be designed to work very efficiently within their wind range.

MAXIMUM SPEED, MINIMUM DISTANCE

Planing boats rarely sail off the wind with an apparent wind angle exceeding 100 degrees. Most often apparent wind is forward of the beam, even when the target is dead downwind. Induced wind is a major component on these boats, which means the apparent wind angle is much sharper. The speed difference between broad reaching/running and close reaching is enormous and much bigger than with heavier, more traditional types of boats. Planing boats will advance quicker, even in a direction dead downwind, if they reach and gybe back and forth, more or less as sailboats normally would do upwind.

VMG

But how much does extra distance does it pay to sail? This is a balancing act and finding the right wind angle in various conditions is precisely the focal point. The goal is of course maximum speed and minimum distance. Come up when you need to build speed, bear away when you can to minimise distance. This is called VMG sailing (Velocity Made Good). VMG is a speed calculation, measuring the actual the speed you approach the target with (up- or downwind) regardless of actual heading or distance covered. Upwind is also referred to as VMG sailing.

DISPLACEMENT BOATS MORE EFFICIENT WITH A SPINNAKER

Heavy displacement boats will often sail on much deeper wind angles. With their more or less fixed maximum hull speed they will benefit from heading directly on target and minimising distance. For this reason the traditional symmetrical spinnaker is a more efficient sail for these vessels. Crews on displacement boats will typically only choose a gennaker to provide a sail that is easier to handle.

REACHING: TRIM AS A SPINNAKER

A lot of the subject matter from the previous pages on spinnaker trim is also perfectly relevant when trimming the gennaker. The gennaker does not have the same amount of trim functions as a spinnaker, but at least on a reach a spinnaker and a gennaker basically work in the same ways. The basic understanding and the same principles can be used, even if you do not have the same tools at your disposal.

DEAD DOWNWIND: DON'T DO IT -- OR TAKE DOWN THE MAINSAIL

If a gennaker is to work properly on a dead run, the mainsail has to be be taken down. Otherwise the gennaker will end up in the lee of the mainsail. Some gennakers are designed with large shoulders, making them better suited for deeper wind angles. As you can see on the picture, they will start to rotate and seek to windward when you come off the wind, especially if you ease the tackline or halyard. This type of gennaker is quite useful for all-round use in traditional cruising boats. Be careful, however, when using the gennaker without the mainsail in strong winds. If it really gets windy, it can be difficult to take it down, as it cannot be brought into the lee of the mainsail when being doused.

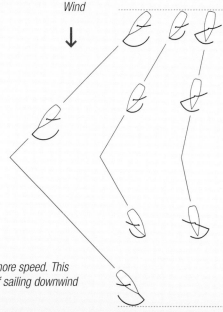

Wind

VMG: More distance, but even more speed. This is often the most efficient way of sailing downwind with a gennaker.

SHEETS, TACKLINE, HALYARD AND BARBERHAUL

Trimming the gennaker is primarily about sheeting and helming correctly. The tackline and halyard pull at either end of the luff, regulating the "shoulder". The sheet and your course determine the angle of attack and if a barberhaul is rigged, this regulates twist and depth.

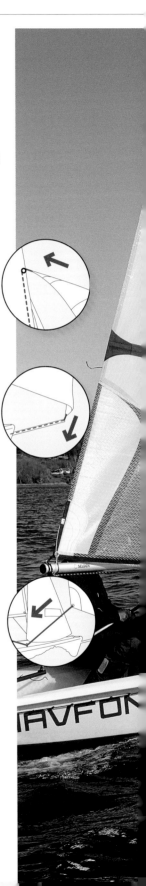

HALYARD

As with the spinnaker, the sail is primarily just hoisted to the top of the halyard. But actually, it could be a good idea to drop it a bit when you want a deeper sail with a rounder shoulder. This is only relevant on very deep wind angles, when you try to come off the wind as much as you can without the gennaker ending up in the lee of the mainsail. Easing the halyard can help the sail to pull to windward, freeing itself from the mainsail wind shadow. It can also work as an alternative to the tackline, if this is fixed. Don't do this in heavy weather, as the sail becomes more unstable.

TACKLINE

The tackline connects the tack of the sail to the bowsprit or bow fitting on the boat. This line can sometimes be adjusted from the cockpit and has more or less the same effect as the foreguy/topping lift on the spinnaker – it controls luff tension. The rule is simple: Broader wind angle, ease the tackline. Sharper wind angle, tighten tackline.

What you want to achieve on a broad reach is a round shoulder, a wide top section and not too much twist. The sail should be helped to windward, away from the wind shadow of the mainsail. Easing out on the tackline will give you just that. If unable to adjust the tackline you can use the halyard instead, with a similar result.

What you want to achieve on a close reach is a flatter, more asymmetrical profile, with the deepest point of the sail more forward and a more open sail with a twisting leech. Tightening the tackline will provide you with that.

For cruising you can comfortably use the gennaker without trimming the tackline. As it is a flying sail and the distance from clew to sheeting point is long, it will adapt fairly well using only the sheet for trimming. A fixed tackline, just long enough to make sure the sail is free of the pulpit will be just fine.

SHEET

Gennaker trim is all about the sheet (in close contact with active helming). The gennaker sheet is trimmed exactly as the spinnaker sheet: Ease out until the luff folds or starts to collapse. A small curl in the luff indicates a good trim.

Sheeting too hard is the most common mistake, so if in doubt, ease out the sheet until the luff starts to curl and pull slightly in. As with the spinnaker, cruisers can lock the sheet and leave it to the helmsman to steer with the luff as a reference.

When it comes to sheeting and helming, there are many similarities between the gennaker and the spinnaker. More on helming and sheeting on page 162.

BARBERHAUL

It is usually unnecessary to use a barberhaul with a gennaker. The sheeting point should be fitted to suit the sail. A gennaker is normally designed for a sheeting point at the back of the boat.

You can however choose to rig a barberhaul on the gennaker sheet just the same as on the spinnaker sheet. The main reason is to be able to reduce twist. It can also be used to stabilise the sail and counteract rolling at low angles in windy conditions.

The effect corresponds (in general) to what you will experience in a headsail when moving the sheeting point forward.

If you fly the clew higher and further off the boat (by easing out or omitting the barberhaul), you will increase twist. Be a little careful with the barberhaul, as it closes the leech and brings the sail closer to the leeward side of the mainsail. This can lead to turbulence. On close reaching it should be completely released.

4 TIPS FOR TRIMMING THE GENNAKER

1   A gennaker should be sheeted as loosely as possible at all times. Sail with a more or less constant light flutter in the top third of the luff.

2   On deep wind angles it helps to fly the clew high and ease the halyard/tackline a little. This helps the sail become full and pulls it to windward, out of the mainsails wind shadow. Be careful when the wind picks up, as this trim makes the sail more unstable. The gennaker could start to oscillate in the waves. Experiment with a looser tackline or halyard and keep an eye on how the sail behaves.

3   Remember to trim the mainsail to avoid interference with the gennaker. The main helps to balance the boat and will control weather or lee helm when sailing with a gennaker. On a close reach it is usually a good idea to trim the boom quite close to the center line and twist it a lot (loosen the kick).

4   On a broad reach, especially in light air, it may be helpful to heel the boat to windward. This helps the gennaker to find free air (to windward).

TIPS AND TRICKS

Rigging and sails in a storm. 170
Storm Sails. 172
Storm strategy 174
Pointing ability 176
Trim guide for good pointing ability 180
Trim guide for light wind 182
Trim guide for medium wind 184
Trim guide to hard wind. 186
Troubleshooting 188

RIGGING AND SAILS IN A STORM

There is every reason to have respect for really rough weather, be it a gale, strong gale or storm force winds. However, with a well prepared boat with an experienced crew there is statistically very little risk. Sails and trim will also affect safety and comfort.

The single most important thing you can do for safety when it comes to storms at sea, is not to be on the water at all when the weather gets really bad. Good weather forecasting is important, along with the will and ability to postpone or cut the trip short, even if it causes troublesome changes to your plans.

The risk of serious problems is also significantly reduced if the boat is robust and in proper condition on the essential points: rudder, keel, through-hull fittings, rigging, sails and engine.

In other words: The most important safety feature is preparation: things you do or don't do while the boat is still in port.

Of course, in spite of good intentions you can find yourself at sea in weather you would not have chosen to be out in. Here are some thoughts on how to handle that situation, with regard to sails and rigging.

SECURED RIG AND APPROPRIATE SAIL PLAN

To handle a storm safely the rig has to be secure and correctly tuned. It really pays off to have a good base trim and sufficient rig tension, as shown in chapter 3.

The next point on the list is to adapt sails to the conditions. Sails used in really bad conditions should be designed and made for them, but once in the situation you will of course have to use whatever you have on board. Reefing to adapt size of sail area is priority number one and this should be done in good time. Even simple and small tasks will become very difficult once the storm hits. All work on the foredeck will be hard and not entirely safe. If there are storm sails on the boat, rig and prepare them as early as possible.

DAMAGE CONTROL
In these conditions trim is not so much about speed. Still, it is important that the boat is moving forward effectively, especially if conditions worsen and there is an opportunity to find shelter, reach port or get out of the weather system. The main thing is to prevent injuries and technical problems. Take care not to overload or damage equipment you will need to sail the boat. This goes of for the whole boat and also when reefing, bending sails and working on the rig.

STABILIZING THE MAST
The base trim and a tight backstay should ensure rig tension in all directions. Do not ease the backstay much on open wind angles, it is more important to stabilise the rig than to optimise sail profile. Check regularly that the mast is straight sideways and has a proper longitudinal pre-bend. If you notice anything strange, ease the sheets immediately to depower the sails and examine the situation more closely. If a shroud breaks on the weather side, the mast will come down very quickly, but if it a fault is detected before it totally gives way, you can save the rig by coming about in a hurry, so the damage ends up on the leeward side. Now the mast can be stabilised with halyards attached to the chainplate or toe rail and tightened up as hard as possible. A broken lower shroud can be "imitated" with a halyard led under the spreaders and preferable also a turn around the mast. This will rule out the use of mainsail but could save the rig.

The mid-section of the mast must never bend backwards, which can happen when the mainsail is deeply reefed and there is not enough tension in the backstay. If possible, the midsection of the mast should bend slightly forward. Any kind of pumping or jerks in the mast must be avoided, if at all possible. Tighten check stay and baby stay if the boat is rigged with these.

REEF AT ALL ANGLES TO THE WIND
Reefed sails should also be trimmed. A simple yet essential check is that halyard and reefing line are tensioned properly. This provides a flat open sail with the deepest point well forward, which is what you need. The sheeting point for the headsail must be adjusted until the sail twists properly

– too little will increase side force, too much could ruin the sail (flogging top/leech).

Also remember to tighten the leech line. If the sail is allowed to vibrate in the leech, the life of the sail will be greatly shortened. The noise from vibrating leeches is also a stress factor for the crew.

It is equally important (but not as obvious) to reef on open wind angles. Lateral forces are less and boat speed greater, so it feels much quieter. This is however a false impression. As conditions worsen, it may become difficult to reef or reduce sail and you could lose control over the boat.

Please feel free to take the mainsail down on open angles – and if not, use a preventer (see page 94).

BALANCING WITH THE MAINSAIL
In rough weather it is important to keep balance in mind. It is usually a good idea to have relatively low pressure in the mainsail compared to the headsail. Upwind, the boat heels a lot, especially in gusts and when hit by waves. You will be able to steer the boat better with the center of effort well forward. The jib sheet should be tight, but allow the sail to twist. Still, the headsail should not be too flat. The optimal sail in storm-force conditions is a very small sail, but with some profile. This provides a wider track and smoother progress. The waves will throw the bow around a lot and precise steering will be difficult. The mainsheet should be set rather loosely but keep a very tight backstay. The mainsail is usually reduced to a tool for balance, not speed. If a gust heels the boat too much, even with released mainsheet, steer into the wind. Ultimately, if needed, ease the jib-sheet too. If this happens a lot, it is a signal that the boat is overpowered for the conditions. Releasing the jib sheet should be a "last resort". If the waves are more on the nose on one tack than the other, point high on the tack where waves are more from the side and lower on the tack where waves are more on the bow.

PRESSURE POINT IN THE SAILS FORWARD
Downwind it is even more important to move the center of effort forward. This will keep the bow down with the wind and reduce the chances of broach and ultimately an involuntary gybe. If the boat is heavy on the helm, do something about it. Sheets are the first place to look. Weather helm is often a signal that the mainsail should be eased out, reefed or taken down.

STORM SAILS

TRYSAIL

A trysail is a storm sail that can be used as an alternative to a fully reefed mainsail. It is very rarely used in practice. If it is going to be of any use, it requires two things: Firstly, that it has been tried out beforehand, so you know the procedures, sheeting point etc. and secondly that it is rigged and prepared before the storm sets in.

STORM JIB

If the boat is rigged with a furling headsail, as most boats are, it could be a challenge to set a storm jib. If you sail on the open sea or want to be prepared for foul weather, it is a good idea to have an inner forestay or cutter stay, where you could hank on a storm jib. Hanks are the most reliable system. There are also storm jibs that can be set "flying", i.e. with the stay integrated in the sail, often with a spectra line. They require a solid mounting point on the foredeck.

Other storm jibs are designed to wrap around the rolled up furler, with a kind of pouch in the luff. Feedback from people who have a lot of experience sailing in rough weather suggests that these storm jibs do not necessarily work that well in practice.

The simplest yet most risky option is to sail with a small piece of furling headsail out. It will work well on open angles and not so well close hauled (see page 88), but the risk is that the furling line could break, or that the system

Boats that are rigged for ocean passages sometimes (if they intend to use a trysail) have a separate track on the mast for the purpose. If setting a trysail requires that the mainsail slides are taken out of the mast track, or if the trysail has to be rigged above the lashed mainsail, then it will be very difficult to rig it in a storm.

becomes overloaded and the whole sail gets torn out of the roll. There are major forces at play. To find yourself with a fully unfurled genoa in storm is a difficult and dangerous situation. If you sail with a reefed furling headsail instead of a storm jib, it is a good idea to secure the furler with a strap to prevent it from rotating.

No matter your choice, you will need to prepare sheeting points fitted for the storm jib.

Storm jib alone is a good sail plan for open wind angles. →

Setting the trysail must be prepared in advance, otherwise it will probably not be manageable. Storm jib is easier, buy you have to know where the correct sheeting point is located and know how the sail is to be set – before the situation arises.

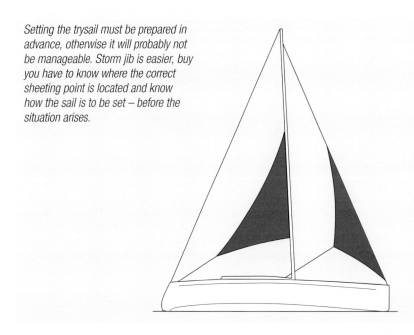

STORM STRATEGY

Sailing with the wind and sea on the beam is not a good idea in a storm. But should you point the bow into the wind and put out a sea anchor – or heave to? Or is it better to run with wind and seas, possibly with a small headsail, trailing a rope to slow you down?

RUNNING WITH THE WEATHER

A lot of seasoned sailors would argue that the best survival strategy for extreme weather is to go with the wind and waves. Sailing with just a small storm jib can be a sensible setup. You could also drop all the sails and sail only on the rig. To run with the weather assumes that there is room enough downwind, however. Experienced sailors are always careful not to sail too close to a weather (lee) shore if there is risk of extreme weather. They will either seek shelter before the weather gets bad, or head far out to sea, with plenty of space to sail with the weather – for several days if necessary.

Avoid placing the boat square to the wind and waves. Breaking waves can turn a boat over if it is beam to. In very high seas you should try to limit your speed down the waves. Long, heavy ropes streaming astern can help and can also curb breaking waves to a certain extent.

A risk when running with the wind and waves is that the helmsman eventually will get exhausted and that sooner or later you will experience a violent broach or knock-down. Moreover, you are sailing many miles in a direction that is not necessarily a good one.

Still, to sail with the weather will be the easiest choice and could be the right decision also in less dramatic circumstances. Even if the situation may not be dangerous, it will often be a good idea to turn around and sail back with the wind, or change the destination to somewhere downwind.

HEAVING TO

Heaving to is an old, traditional way to survive harsh weather. Modern boats are not always designed and balanced in a way that makes the technique suitable and the method is rarely used these days. But especially with a traditional, long-keeled boat it may be relevant to heave to and wait for better weather. You should try it out in controlled conditions, to know how your boat reacts. This is how it is done: When the boat is sailing close-hauled (upwind) you initiate a normal tack, but leave the jib sheet attached so that the jib is backed. The mainsheet is loosened and adjusted so the boat balances, but without the mainsail contributing much to forward progress. The rudder is hard over, steering the boat into the wind. The backed jib now forces the boat to leeward, while the helm pushes the boat to windward.

The idea is that these two forces will counteract each other and keep the boat in balance. When this works as intended, the boat should be in balance, with the bow pointing in a 45 degree angle towards the waves and wind, with hardly any speed ahead, but also without drifting much. The boat will now deal with the waves relatively well. The rudder may be locked in this position, so the whole crew can theoretically remain below deck until the weather improves, maintaining a proper lookout. In extreme weather, the boat might be knocked down by the forces in the backed headsail, but never the less; many have safely weathered a storm this way.

SEA ANCHOR

A drogue or sea anchor helps to keep the bow towards the waves. The rope should be long and with stretch and the sea anchor should preferably be in step with the waves, so the boat and drogue are on top of a wave at the same time.

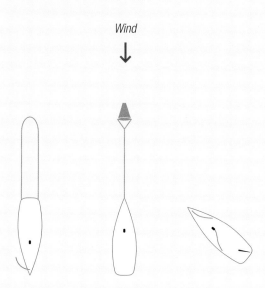

Wind

Left: Dragging lines ("warp") behind the boat is helpful when running with the weather. It reduces speed down the waves and helps keep the stern against the wind.

Middle: Sea anchor keeps the bow up against the waves and reduces drift. Sea anchor can also be set from the stern, but in this case the helm must be secured very well and the hatch to the companionway has to be secured. All sails must be bent on with extra care.

Right: Heaving to is a well proven storm technique recommended by a lot of sailors with practical experience of storms on the ocean. Traditionally designed boats will normally respond better than modern designs, but everyone should experiment with their own boat and find out if and how it best works.

4 TIPS FOR SAFETY IN ROUGH WEATHER

1 Stay on board. In other words: Hold on! Be aware and take care when you move about on deck. Use a harness when conditions demand it.

2 Use a life vest, preferably one with a personal emergency beacon, so that you can easily be found if you should fall in the water.

3 Be aware of the boom when the boat sails downwind, especially in high seas. Use a preventer, make sure the helmsman is focused and that the crew knows what an involuntary gybe is. Avoid a dead downwind run, especially with any kind of mainsail up.

4 Check if the boat is taking on water, either through leaky or defective through-hull fittings, piping systems, deck hatches or elsewhere. Don't forget to concentrate on navigation!

POINTING ABILITY

Good pointing ability! Easy to say – but what actually gives you good pointing ability? What factors are really involved?

If a boat is to sail efficiently to windward, it must produce minimum drag. Only then can we trim the sails very flat and to the centre line, maintaining consistent speed, assuring good flows around the sails, keel and rudder blade.

TWO "CASE-STORIES":

Let's be specific and compare two boats: The first is a light boat with high stability, low freeboard, low superstructure, tall slim rig and narrow, deep keel and rudder. Such a boat creates little resistance and a lot of forward thrust and can therefore sail with very flat sails, trimmed to a small angle of attack. The pointing ability can be very good. The second boat is heavy with low stability, high freeboard, large superstructure, low rig and a long, shallow keel. This boat does not create much forward thrust but has a lot of resistance, both through the air and through the water. The sails have to be deeper in order to create the necessary force for progress. The attack angle of the sail must also be larger, to ensure that a greater proportion of the lift pulls the boat forward. The pointing ability of such a boat can never be good.

THE TORTOISE AND THE HARE

There are two consequences of the examples above: The first is that you not only trim for the circumstances, but also within the potential of the boat you actually have. The second is that boats have varying potential when it comes to pointing ability and you can't overcome all of these differences without resorting to buying another boat. You cannot turn a tortoise into a hare – but as we shall see, you can turn it into a speedy tortoise.

SPEED + POINTING ABILITY = TRUE

Pointing is not only about pointing the bow in a direction closer to the wind. That will not necessarily bring the boat more efficiently to a position in a windward direction. A boat trimmed with extremely flat sails and a very small angle of attack can be steered at an angle very close to the wind. But this boat will be very slow. The loss of speed could easily end up costing more than the saved distance. Furthermore, when speed drops, drift will increase. You could end up sailing more or less the same course over the ground (COG) as a boat that moves faster on a lower heading. The difference is that you will move slower and more sideways. A fast boat will generate more lift and build the potential for a better preserved pointing ability. Speed and pointing are in other words related. You need to be fast to point well.

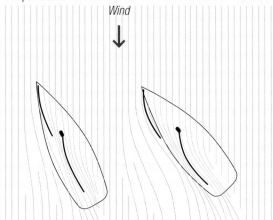

Wind

If you point very high, like the boat on the left, the sails have to be flat and the angle of attack small. In this situation the sails cannot produce as much lift and forward thrust will be reduced. When speed drops, drift will increase. The boat on the right sails with fuller sails and a larger angle of attack, i.e. lower and faster. Sailed distance upwind will be considerably longer. A middle road is often the best.

THE BOAT IS PART OF THE PICTURE

Pointing ability is not just about sail trim. It is about the whole boat, including what is happening below the surface. Under water the hull, keel and rudder also have an influence. The underwater configuration creates both drag and lift, but unless it is a foiling boat, drag will overcome lift by a lot. Luckily, the opposite is true when it comes to the sails, as they produce much more lift than drag. In both cases we are always looking for maximum lift and minimum drag.

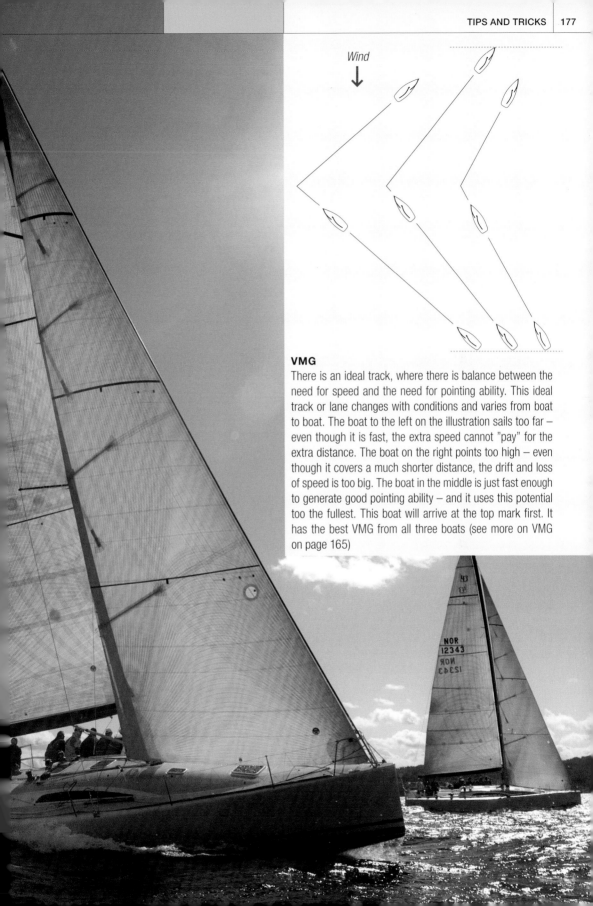

Wind ↓

VMG

There is an ideal track, where there is balance between the need for speed and the need for pointing ability. This ideal track or lane changes with conditions and varies from boat to boat. The boat to the left on the illustration sails too far – even though it is fast, the extra speed cannot "pay" for the extra distance. The boat on the right points too high – even though it covers a much shorter distance, the drift and loss of speed is too big. The boat in the middle is just fast enough to generate good pointing ability – and it uses this potential too the fullest. This boat will arrive at the top mark first. It has the best VMG from all three boats (see more on VMG on page 165)

POINTING TOO HIGH (PINCHING)

Most people have probably tried to get up to a mark, a navigational obstruction or something else without tacking, but have been tracking a little low. It would seem obvious to steer a bit higher than the optimum course, in an attempt to save two tacks and a little extra distance. This is known as pinching. You can manage a few degrees extra if the sails are adapted, but if you exceed these few degrees boat speed drops very quickly. The drift increases, the boat goes slower and slower and CMG (course made good) is worse than it would have been if you had continued with full speed. It is a strategy that only works if it is only a matter of covering a few boat lengths before falling off again and regaining speed.

POINTING TOO LOW

When you come off the wind, you build up speed quickly. However, when coming off the wind with close-hauled sheets, heeling will soon increase more than speed. Your VMG (velocity made good) will not be particularly good. You may be fast enough, but the angle against the wind is now poor. You will simply sail too far – further than the slight speed increase can offset.

THE GROOVE

There is quite a narrow groove or lane to operate in, where you can either adapt the rigging and sails for some extra speed or benefit from a little extra pointing ability. When racing you are often forced to switch between "pointing mode" and "speed mode", especially after the start, when you are pushed by other competitors, either to windward or leeward. You can trim and go for a little extra pointing ability or a little extra speed to get out of a close quarter situation, but there is not a lot of leeway. Sail higher or lower than this defined groove and you will lose either way. In a traditional displacement boat we are talking about 6-8 degrees.

OLD SAILS – POOR POINTING ABILITY

Sails will stretch. This applies in particular to dacron sails. They stretch under load caused by wind pressure and far more when close hauled than on open angles. When sails have been used over time, they will begin to stretch more. The aft part of the sail is loaded the most and also stretches the most. Two unfortunate things now happen at the same time: Fullness increases and the deepest point moves backwards. The sail will produce more drag, more lateral forces and less lift, meaning less progress.

You can do a lot by trimming. Tighten the halyard, the cunningham or outhaul, tighten sheets and backstay etc. But when the sail gets old and the aft part of the sail starts to "bag", the problem is that you can no longer tighten the sheet as much as you would like, without deforming the sail. In order to provide the necessary twist in the sail you will have to ease the sail out slightly. But this will cost you pointing ability. The only thing that really helps in this situation is a new sail.

CAN YOU SHEET THE HEADSAIL FAR ENOUGH IN?

One of the decisive factors for a boat's potential pointing ability is the practical possibilities there are for making the sheeting angles small enough. With older, long-keeled boats with short masts there is no need to worry; their headsail should most likely be sheeted out by the toerail anyway, due to their limited potential pointing ability. Performance boats with tall rigs and fin keel will however benefit from being able to move the sheeting point closer to the center line. An inhaul or a barberhaul (see page 136) will be a good aid. The best solution is probably a rig with several short spreaders and chainplates mounted well on deck. This allows for a sheeting track mounted nearer the center line. Remember that the possibility for a tight sheeting angle does not mean that it is always a good idea to use it. You must be able to generate speed. Speed creates pointing ability!

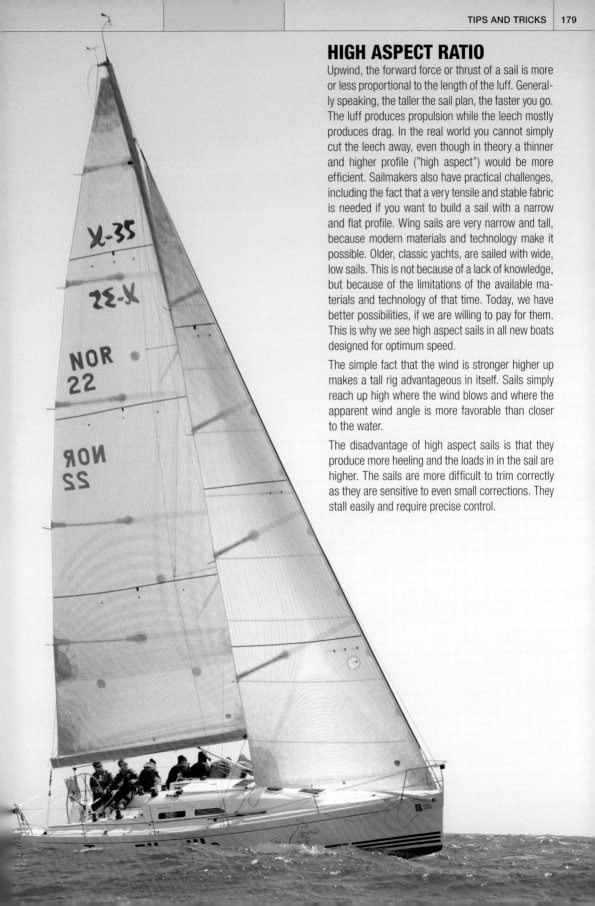

HIGH ASPECT RATIO

Upwind, the forward force or thrust of a sail is more or less proportional to the length of the luff. Generally speaking, the taller the sail plan, the faster you go. The luff produces propulsion while the leech mostly produces drag. In the real world you cannot simply cut the leech away, even though in theory a thinner and higher profile ("high aspect") would be more efficient. Sailmakers also have practical challenges, including the fact that a very tensile and stable fabric is needed if you want to build a sail with a narrow and flat profile. Wing sails are very narrow and tall, because modern materials and technology make it possible. Older, classic yachts, are sailed with wide, low sails. This is not because of a lack of knowledge, but because of the limitations of the available materials and technology of that time. Today, we have better possibilities, if we are willing to pay for them. This is why we see high aspect sails in all new boats designed for optimum speed.

The simple fact that the wind is stronger higher up makes a tall rig advantageous in itself. Sails simply reach up high where the wind blows and where the apparent wind angle is more favorable than closer to the water.

The disadvantage of high aspect sails is that they produce more heeling and the loads in in the sail are higher. The sails are more difficult to trim correctly as they are sensitive to even small corrections. They stall easily and require precise control.

TRIM GUIDE FOR GOOD POINTING ABILITY

1   Minimal resistance, both through the water and air

The boat must be able to keep a good pace, even with relatively small forward propulsion (a large proportion of the lift from the sails pulls sideways or backwards, when the boat points very high). Maintaining speed requires minimal resistance, both above and below the waterline. Some boats have better potential than others. Boats with good pointing ability are lightweight boats or boats with good stability. Small wetted surface, deep keel, low freeboard and superstructure, a high rig, long forestay and high aspect ratio sail (high, narrow sail plan) are all beneficial features.

It helps to have a clean hull and minimal equipment to create windage and drag above deck.

2   Sheeting angle close to the centre line

Having the sheeting track for the headsail close to the centre line offers potential for better pointing ability. If the track is well out on the side deck, you can use an inhaul on the sheet to pull it inward. On the mainsail, the traveller should often be set to windward of the centerline. The boom should be on the centre line on most boats and both sails sheeted in tightly — but make sure speed does not drop more than a fraction. You need good pressure in the mainsail to maintain good pointing ability, at least in light and medium winds. The leech must not open too much.

3   Sails with a flat entry

A flat entry provides better pointing ability. In practice, you create a flat entry with relatively little tension in the halyard on both sails. This can't be a priority in strong winds however, because heeling must be regulated and this requires a tight halyard. Backstay tension is also a big factor when trimming for a flatter entry.

4   Flat sails with the deepest point relatively far aft.

Flat sails are basically achieved by pulling out luff curve — that is, having a tight forestay and a curved mast. The backstay or running backstays (and the mainsheet) must be set relatively hard, to make sure the rig tension longitudinally is high enough. To pull the foot of the mainsail flat, tighten the outhaul.

5 Limit twist

The leech must not open too much. In the headsail twist is regulated with the jib car position and the jib sheet. In the mainsail, twist is regulated with the traveller and the mainsheet. The headsail sheeting point should be trimmed precisely until the headsail twists exactly the right amount when the sheet is tight. The traveller should be trimmed until the boom is situated on the centerline when the sheet is tight enough to provide optimal mainsail twist. Use telltales to find the right twist. When heeling exceeds optimal heel angle, ease the traveller until heeling is right.

6 Limit heeling

Adjust heeling to the optimum angle for the boat. Try to trim the sails flatter and twist them slightly more before eventually reefing. If the boat doesn't heel enough you are not exploiting the boat's potential. If the boat heels too much, speed drops and drift increases.

7 Well balanced center of effort

There should be a slight weather helm present – the boat should seek to windward, but without insisting. This creates lift across the rudder and to some degree across the keel as well and makes it easier for the helmsman to steer precisely. The boat should still be close to perfectly balanced. The preferred helm pressure is easily achieved by adjusting the mainsail angle of attack (move the traveller to leeward to ease helm pressure, or to windward to increase helm pressure). When pointing ability is lacking, a tighter mainsheet is often the answer. If you need to reduce weather helm, ease a little on the headsail sheet or try dropping the backstay a bit. This will initially cost you some pointing ability, but might be what is needed in order to keep up your speed.

8 A helmsman with a precise touch

As soon as the sails are trimmed correctly a good helmsman becomes a very important issue. A good helmsman is able to keep the boat "on track" at all time, i.e. within the few degrees of the defining groove that perfectly matches the sail's angle of attack. With sails trimmed to maximum pointing ability, this is no mean feat. A flat entrance is difficult to steer by! Try to make windward telltales in the headsail barely flicker. If you manage to keep this balance over time, windward telltales will start pointing slightly upward. This is a good sign.

Having the crew working on the trim (primarily with the sheets) even with small shifts in wind strength is also a great help.

TRIM GUIDE FOR LIGHT WIND

1   Looser halyard

Ease halyard tension in both the headsail and mainsail. In very light winds it doesn't matter if fine, horizontal stripes appear in the sail's forward section. This makes the sail deeper, the deepest point moves aft and helps to close the leech, giving a powerful sail.

2   Looser backstay

Release the backstay to straighten the mast. Watch the forward part of the mainsail. There should be a nice, smooth, entry to the sail. Keep an eye on the forward part of the headsail as well. This could do well with a little sag, that is, a somewhat loose tension in the forestay. The entry to the headsail can be relatively round, but do not over-do it. Adjust the backstay until the profile in the forward part of both sails looks good and make sure the mast is not straightened so far that the sails pull too much fabric forward towards the luff. This is a fine (and important) balance.

3   Looser outhaul

Make sure the outhaul is a little looser than usual. On most boats the bottom of the mainsail should still be relatively flat. When there are waves it should be looser than on flat water.

4   Looser headsail sheet

Tighten the sheet until the foot retains a slight curve. The sail should be a little deeper than usual. Keep an eye on the top telltale in the leech, which should be streaming about 80% of the time and check the gap between the sail and lower spreaders. This should be about the width of a fist. Adjust the jib sheet car until this is the case. On flat water, some boats will benefit from a barberhaul or inhaul to move the sheeting angle slightly closer to the boat's centerline. This will improve pointing ability. A very common mistake in light air is to sail with the sheet to tight and not enough twist in the sail.

6   Looser mainsheet

Tighten the mainsheet until the leech has a reasonable amount of twist. Keep an eye on the top telltale in the leech. It should be streaming about 80% of the time. Also check the angle between the boom and the top batten. They should be close to parallel. The top batten must not "hook" (point to windward). A very common mistake in light air is to sail with the sheet to tight and not enough twist in the sail.

5   Traveller to windward

Trim the traveller to windward until the boom ends up close to the centre of the boat. The kick should be completely free.

DOWNWIND

Follow the trim guide for close hauled sailing on the opposite page, but set everything even looser.

Set the kick gently, just enough to regulate the twist in the mainsail. Do not close the sail too much.

Pay attention to the sheets. On a reach they should be checked frequently. Make sure they are not too tight.

If you are using a spinnaker or gennaker, the headsail should be taken down or furled, although it could possibly be kept up on a close reach.

More about spinnaker trim on page 152-163.

Gennaker trim: page 164-168.

TRIM GUIDE FOR MEDIUM WIND

1   Slightly tighter halyard

Slightly more tension in both headsail and mainsail halyards. There should be no horizontal stripes in the foremost part of the sail, but close. The halyard should not be tightened so hard that the sail creates a vertical fold along the luff (when sheeted and on actual heading). This is a way of helping sails to maintain the shape they were designed for.

2   Slightly tighter backstay

Tighten the backstay until the mast has a bend that makes the luff curve in the mainsail look right. Check the forward part of the mainsail. There should be a nice smooth entry to the sail. Pay attention to the forward part of the headsail as well. Sag will be reduced, but the forestay should (generally) not be tightened to the maximum. The entry to the headsail can be relatively flat, but adjusted for waves (reduce sag in flat water, more sag in waves). Adjust the backstay until the profile in the forward part of both sails looks reasonable. Make sure that the mast does not bend so much that valuable power is lost. If the boat heels too much, tighten more on the backstay. If you can use more power, ease the backstay.

3   Slightly tighter outhaul

Tighten the outhaul. On most boats the bottom section of the mainsail should now be quite flat. In waves it should be looser than on flat water.

4   Slightly more tension in headsail sheets

Tighten the headsail sheet until the sail's foot has a slight curve. The sail should be a little flatter than in light winds. Watch the top telltale in the leech, which should be streaming about 80% of the time and check the distance between the sail and lower spreaders. This will usually be about the size of a fist. Adjust the sheet car until that is the case. On flat water, some boats will benefit from a barberhaul or inhaul to move the sheet angle slightly closer to the boat's centerline. This will improve pointing ability.

5   Slightly more tension in the mainsheet

Tighten the mainsheet until the leech has a reasonable twist. It should be fairly tight – tighter than in light winds and most often also tighter than in strong winds. Keep an eye on the top telltale in the leech. This top telltale should be able to stream about 80% of the time.

6   The traveller close to the middle

Trim the traveller until the boom is in the middle of the boat, or even slightly to windward. In medium wind the traveller can be quite close to the centerline. If the boat heels too much or develops too much weather helm, drop the traveller slightly to leeward until the problem is resolved. This means less power in the mainsail, so be careful to do this no more than necessary. Pull it back up whenever you can.

DOWNWIND

Follow the trim guide for sailing close hauled on the opposite page, but with less tension.

Set the kick to adjust mainsail twist. Do not close the sail too much, but make sure you control the twist.

Trim sheets frequently. Make sure they not end up too tight. If in doubt, ease out.

If you are using a spinnaker or gennaker, the headsail should be taken down or furled, although the headsail could possibly be kept up on a close reach.

More about spinnaker trim on page 152-163.

Gennaker trim: page 164-168.

TRIM GUIDE FOR HARD WIND

1 Tight halyard

Tighten the halyard really hard, both for the headsail and the mainsail. Now we need to release excess energy in the sails and counteract a lot of deformation due to wind pressure. Tensioning the halyard a lot will pull depth forward, flatten the sail and open the leech. This is very efficient and just what you need!

2 Tight backstay

Tighten the backstay hard. How much mast bend you need depends on weather and rig type: A fractional rig is usually trimmed with greater mast bend than a masthead rig. Most boats will benefit from a lot of longitudinal mast bend when the wind is up. The mast curve will flatten the top and middle portion of the mainsail and open the leech. At the same time depth is pulled aft, but this can be offset by halyard tension. The forestay should generally be as tight as possible, to minimise sag and give the headsail the same treatment as the mainsail (luff curve pulled out, flatter and more open sail).

If the boat heels too much, try to tighten the backstay even more. If the boat has trouble keeping pace in rough seas, try easing slightly on the backstay and sail a little lower (remember to adjust sheets).

3 Tighten the outhaul

Tighten the outhaul to the maximum. The bottom of mainsail should be flat.

4 Tight jib sheet

Tighten the jib sheet hard until the foot of the sail only has a tiny curve. The sail should be slightly flatter than in medium winds, but make sure you do not pull all the profile out of the lower section of the sail. This is especially important in a high aspect headsail, which must not be flattened too much in the lower part. You need a certain pressure and momentum at the head of the boat for speed and balance. The headsail can twist a bit more than usual and it might be an idea to move the sheeting point slightly aft. It is probably not a good idea to use the barberhaul or inhaul, as this will increases lateral force. Hard wind usually means larger waves and that makes it necessary to sail a little bit lower.

5 Tight mainsheet

Tighten the mainsheet until you are happy with balance and heel and ignore the top telltale. The sail is usually sheeted looser than in medium wind and it is often necessary to trim actively in the gusts. Dealing with gusts can also be managed with the traveller (see more about this on page 150). If the leech flogs a lot of the time, it is a sign that the boat is overpowered or not trimmed correctly.

6   Traveller to leeward

Trim the traveller until the boom comes as close to the boat's centre line as possible, when the mainsail is sheeted with an acceptable twist. The degree of heel will determine how far up it can come. If you have too much weather helm, drop the traveller slightly until the problem is resolved. In gusts of wind you can either ease the sheet or the traveller, but remember to adjust back in between the gusts. Some luffing of the mainsail is OK.

7   Reef when necessary

Upwind you can trim yourself out of a lot of trouble. When the wind speed reaches a certain point, however, it is time to put in a reef. This point varies widely, but as skipper or boat owner you should know before you have reached it. When cruising you should not be overpowered in gusts. In racing, sailors tend to have the opposite attitude: The boat should not be underpowered between gusts. More on reefing on page 86.

8   Use the helm actively

In strong wind, try to keep a constant and restricted angle of heel. Wind and waves make it necessary to helm actively and to concentrate. In these conditions, heeling is a very good reference for helming. Steer to windward when the boat heels more; steer to leeward when the boat heels less. This way you can maintain good speed and be reasonably comfortable.

DOWNWIND

Follow trim guide for close hauled sailing on the opposite page, but set everything a bit looser.

Adjust kick until twist in the mainsail is regulated and controlled. Close reaching: A lot of twist. This helps balance and reduces heeling. Broad reaching/running: Less twist. Too much twist makes the boat unstable.

Check sheets frequently. When broad reaching, make sure they are not set too tight. Downwind at deep angles, do not ease the sheets (especially mainsheet) too much, as this increases the risk of a broach and involuntary gybe.

Only use a spinnaker or gennaker if the crew is experienced. The headsail must be taken down or rolled in.
More about spinnaker trim on page 152-163. Gennaker trim: page 164-168.

Reef or take down sail in good time. Carrying too much sail on open wind angles increases the risk of problems, such as overpowering the rigging, sails and rudder, or losing control and broaching. On deep angles it is a good idea to take the mainsail down and just use the headsail. This will provide a safer and more stable situation. If the mainsail is up, use a preventer (see page 94).

TROUBLESHOOTING

UPWIND

PROBLEM	POSSIBLE CAUSE CHECKLIST	POSSIBLE SOLUTION
POOR SPEED	Is there enough sail up?	Remove any reef, roll out more of the headsail or set a larger headsail.
	Are the sails sheeted for optimal angle of attack?	Tighten or ease sheets until the angle of attack is optimal. Ease the sheet until the sail luffs and then tighten slightly.
	Are the sails too flat?	Ease the halyard, outhaul or backstay until the sails regain power.
	Is the depth of the sail placed too far back – or too far forward?	Adjust the halyard, outhaul and backstay until the deepest point is positioned correctly (40-50%, further ahead in headsail than mainsail).
	Are the sails twisting too much – or too little?	Adjust the jib sheet car and jib sheet as well as the traveller and mainsheet, until the twist in the sails is correct.
POOR POINTING ABILITY	Is the speed OK? (no speed, no pointing ability.)	Make sure that speed is good. Follow the checklist above.
	Is the deepest point in the sails too far forward?	Ease the halyard (one or both), until the entry of the sail is flatter and the deepest point is placed further back. The backstay may need to be tightened.
	Are the sails sheeted close enough to the boat's centerline?	Move the traveller to windward and use any barberhaul or inhaul to move the headsail sheeting angle closer to the centerline. Tighten the halyard, outhaul or backstay until the sails are flatter.
	Are the sails too deep?	Tighten backstay, outhaul and halyard. Tighten jib sheet. Experiment with mainsheet tension.
	Are the sails twisting too much?	Reduce twist by moving the jibsheet car forward (or alternatively, tighten the jibsheet). Also tighten main sheet. Move the traveller to leeward if more mainsheet tension moves the boom above the centre line, or too far to windward.
POOR BALANCE (WEATHER HELM)	Is the boat heeling too much?	Tighten halyard, outhaul or backstay until the sails are flatter. If the sails are flat enough, tighten the jib sheet and ease the mainsheet. If this is already done, ease the traveller to leeward.
	Is the headsail too small relative to the mainsail?	Take in a reef in the mainsail, or (if the boat heels less than optimal heel angle) increase the headsail area by furling it further out. Change to a larger headsail, if the boat is rigged with headfoil or hanks.
	Is the mainsheet too tight?	Ease the mainsheet.
	Is the traveller set too far to windward?	Drop the traveller to leeward.
	Is the mainsail too deep, too closed, or trimmed with the deepest point too far back?	Tighten the halyard and outhaul until mainsail is flatter. Be careful with backstay tension here (unless it is windy) as this will reduce power in the headsail.
	Is the mast trimmed too far aft?	Trim the masthead further forward. This is done by reducing the length of the forestay and (if rigged with swept spreaders) adjusting cap shrouds and lower shrouds. This cannot be done underway and is mostly relevant for class boats and dinghies.

DOWNWIND

PROBLEM	POSSIBLE CAUSE CHECKLIST	POSSIBLE SOLUTION
POOR SPEED	Not enough sail for the wind conditions?	Hoist more sail. Shake out any reefs, unfurl the headsail fully or hoist a larger headsail. Hoist a gennaker or spinnaker.
	Is the headsail covered by the mainsail?	If the heading is low enough, pole out the headsail using a spinnaker pole. If reaching, point slightly higher, to make sure apparent wind reaches the headsail without being interfered by the mainsail. Consider dropping or furling the headsail and setting a spinnaker or gennaker.
	Are the sails sheeted in too tightly?	Adjust the sheets until the angle of attack is optimal.
	Are the sails twisting too much – or too little?	Adjust the kick and move the jib sheet car until the sails twist correctly.
	Are the sails trimmed too flat?	Ease backstay, halyard and outhaul. When sailing with a spinnaker or gennaker: Check the trim instructions page 152-168.
	When sailing with a spinnaker or gennaker: Is the sail trimmed correctly to the wind angle?	Check the trim instructions page 152-168.
POOR BALANCE	Is there too much power in the mainsail – or too little power in the headsail?	If the boat has weather helm: Ease the kick, ease out on the main sheet or reef the mainsail. Put out a larger headsail or move the jib sheet car forward. See check list for poor balance upwind, it is largely usable on any wind angle. If the boat has lee helm: Tighten the kick, tighten the mainsheet, remove any reef in the mainsail (most effective on a close reach). Reduce the headsail area or move the jib sheet car aft.
	When sailing with a spinnaker or gennaker: Is the sheet too tight?	Ease the sheet until the sail starts to luff and tighten slightly.
POOR CONTROL (UNSTABLE BOAT, DIFFICULT TO CONTROL)	Is the sail area too large for the wind strength?	Reduce sail. Take down any spinnaker or gennaker, reef the mainsail or take it down, furl the headsail, or change to smaller headsail. Remember to maintain balance when you adjust sail area – see checklist for poor balance (above).
	Are you sailing too low (dead downwind) – or too high (too much lateral force and heeling)?	Sail slightly higher or lower (depending on the wind angle), to achieve better stability. In heavy weather, the most stable lane is most often between 140 and 160 degrees apparent wind. In lighter winds the ideal track is a lot closer to the wind. Adjust sheets to suit the course.
	Do the sails twist too much?	Tighten the kick. Try moving the jib sheet car forward. When sailing with a spinnaker or gennaker: Check trim manual on page 152-168.

INDEX

Aluminium 25
Angle of attack 104
Apparent wind 58
Baby stay 9, 22, 125
Backstay 21, 125
Balance........................... 103, 112
Barberhaul 136, 160, 166
Batten...................................... 63
Cap shroud 18
Carbon..................................... 26
Check stay 9, 22
Clew 62
Cunningham 140
Cutter stay 9, 22
Depth.............................. 103, 114
Douse 73
Downwind 56, 164
Drag 53
Fathead 70
Foot.................................. 62, 63
Forestay 20, 126
Fractional rig 14
Fullness 103, 112
Furling headsai......................... l80
Furling mainsail 83
Furling reef.............................. 80
Furling systems 80
Gennaker halyard 166
Gennaker sheet 166
Genoa sheet 130
Guy 152, 160
Halyard 132, 139

Handling................................... 75
Headboard 63
Heel angle 106
High aspect ratio 179
Hoist sail 73
Induced wind........................... 58
Intermediate 19
Jib hank 79
Kick 142
Laminar airflow 54, 104
Leech................................. 62, 63
Lift.. 53
Lower shroud 19
Luff.................................... 62, 63
Luff curve.............. 62, 63, 69, 79
Maintenance 48, 96
Mast curve 127
Mast bend............................. 127
Masthead rig 14
Outhaul 141
Pointing ability 176, 178, 180
Reach 56, 164
Reef....................................... 87
Reefing 63
Rig... 8
Rig tension 42
Rig trim 36
Rigging.................................... 33
Rod... 28
Rope....................................... 30
Running backstays......... 9, 22, 128
Running rigging 30

Sag 126
Sail 51, 61
Sail profile 68, 114
Sail trim 101, 121
Sheet 146, 152, 166
Sheeting point 134
Shroud.................................... 17
Sock (spinnaker/gennaker) 84
Speed 176, 188
Spinnaker downhaul................ 158
Spinnaker pole....................... 158
Spreader 17
Stalling.................................... 54
Standing rigging 28
Stay.. 17
Storm................................... 170
Storm strategy........................174
Storm sails 172
Tack.. 62
Tackline................................. 166
Telltale 122
Topping lift 158
Traveller 144
Trim functions........................ 121
True wind 58
Tweaker 136
Twist...............108, 142, 144, 146
Upwind.......................... 56, 176
VMG 164
Vortex 110
Wire.. 28